CONSTRUCTION MATERIALS AND ENGINEERING

CONSTRUCTION PROJECTS

IMPROVEMENT STRATEGIES, QUALITY MANAGEMENT AND POTENTIAL CHALLENGES

CONSTRUCTION MATERIALS AND ENGINEERING

Additional books in this series can be found on Nova's website under the Series tab.

Additional e-books in this series can be found on Nova's website under the eBooks tab.

CONSTRUCTION MATERIALS AND ENGINEERING

CONSTRUCTION PROJECTS

IMPROVEMENT STRATEGIES, QUALITY MANAGEMENT AND POTENTIAL CHALLENGES

KIMBERLY HALL
EDITOR

nova publishers
New York

Library of Congress Cataloging-in-Publication Data

ISBN: 978-1-53610-742-5

Published by Nova Science Publishers, Inc. † New York

CONTENTS

PREFACE

The sustainability of the construction industry is a matter of pressing concern. Construction activities pose a significant burden on the environment. This book reviews different improvement strategies for construction projects. It also review management models and discusses challenges that arise in construction projects.

Chapter 1 – This chapter concerns the need to evaluate the reliability of structural elements produced with recycled aggregates concrete (RAC) in order to address some designers' scepticism towards the use of this eco-friendly material. The current knowledge on RAC's behaviour demonstrates its viability for structural purposes. However, of the investigations performed so far very few are related to a fundamental aspect towards RAC world-wide application as a structural material: structural codes have a probabilistic basis. After briefly presenting the state-of-the-art knowledge on the material properties and structural behaviour of RAC, the limitations of the current knowledge are debated. Afterwards, an introduction to structural codification is presented, as well as the fundamentals of reliability analysis. Examples of code verifications are contextualized with their underlying assumptions and the information necessary for code calibration is discussed. The role of reliability in the calibration of structural codes is shown, common techniques for reliability calculations are briefly explained, and relevant references in the area are cited for the readers' perusal. Having established how structural codes are calibrated, the state-of-the-art on the probabilistic and statistical knowledge of RAC properties is reviewed. The implications of the very reduced number of studies on this area are discussed and the need to conduct further studies is emphasized. Afterwards, investigations that used reliability analysis to calibrate partial safety factors applicable to RAC are reviewed. The methodology of each investigation is presented, the experimental tests that led to the definition of the probabilistic information of the RAC's parameters are described and the need to have a wide range of data coming from different RAC compositions and aggregate sources is debated. This chapter finishes by contextualizing the current knowledge on RAC properties with the necessary information for code calibration procedures. The relevance of a code proposal towards RAC affirmation as a structural material is highlighted, as well as the requirements of such code. Suggestions for future studies are made.

Chapter 2 – This research analyses how business processes of a company are related to the processes of projects, as the latter generally concentrate on the optimal joint to initiate, plan, execute, control and close a project, leaving aside the collaborative learning environment, which can benefit the development of the projects, and wasting technological tools provided by the use and appropriation of ICTs (Information and Communication Technologies) for knowledge sharing and transfer. Therefore, this research proposes to deploy a model of Knowledge Management for Project Management $KM+PM_{TIC}$ in order to facilitate best practices in the implementation of future projects, through the use of techniques and tools that facilitate knowledge creation, sharing and transfer, and supported by information technologies and communications. The final aim is to optimize the processes used in the development of projects from beginning to end, satisfying time restriction, costs, scope, quality, risk and use of resources, and motivating the project team members to work in collaborative environments that optimize the organization, letting them benefit from the experiences, lessons learned, best practices, generate new knowledge and increase the percentage of successfully completing projects.

Chapter 3 – Bitcoin, the first cryptocurrency in the world has been in use since 2009. Blockchain is a technology that underlies the use of bitcoins. In recent times, blockchain technology is being touted as a promising tool which can transform many business operations. Indeed, it has been described as the next big thing after the emergence of the internet. This study examines the blockchain technology with the view to informing the construction industry on the potential of this technology to enhance value delivery. The study explores the potential uses, benefits, challenges and implications of the blockchain technology for the construction industry. Additionally, it provides future research areas for the leverage of the blockchain technology in project delivery and other construction-focused activities.

Chapter 4 – The production and use of massive data and information are key intrinsic characteristics of construction value delivery. Therefore, the use of 'big data' techniques/tools can potentially reveal better insights for construction value delivery than the traditional methods applied in dealing with construction data and information. Yet, the use of 'big data' techniques/tools is a recent concept, and insufficient studies exist in the literature to guide research and practice in the use of such technologies in value delivery in the construction industry. This study fills part of the gap in extant knowledge. The study extends knowledge by exploring the benefits, challenges and the potential uses of 'big data' techniques and visualization tools in construction value delivery. Additionally, the study provides recommendations for practice and directions for further research. These contributions can help harness the benefits and deal with the threats in the use of this novel technological concept (i.e., 'big data') in construction value delivery.

Chapter 5 – To cope with the growing population over the years, the Hong Kong government spends more than HK$70 billion in infrastructural and institutional developments annually. Being the largest client in the construction industry, the government strives to complete the public projects in a timely and cost effective manner. However, the lowest-bid tendering method and the increasingly complex project nature raise a large number of construction disputes in the public projects. In 2009, the government piloted the use of NEC3 contract (a form of relational contract) in a few civil engineering projects, attempting to alleviate the problem of construction disputes by establishing partnership with the contractors. The projects were satisfactorily completed, with time and cost savings to the government. More importantly, the number of disputes in each project was substantially reduced as most of the construction or interfacing problems were resolved once they became apparent. Although building partnerships at work has been promoted for a long time, distrust between the project parties is deep-rooted. The conventional way of non-contractual partnership brings very little improvement in dispute mitigation. Successes in the NEC3 projects provide valuable information for us to design and implement relational contracts in order to minimise construction disputes. Data were collected from a questionnaire survey and follow-up interviews with 83 project participants who were involved in three projects using NEC3 contracts. The key attributes that promote partnership in a relational contract were identified and the effectiveness of partnership developed through relational contract attributes was cross-examined with Macneil's relational contract norms. Project stakeholders can use the recommendations to promote collaboration and minimise disputes through relational contracting.

In: Construction Projects
Editor: Kimberly Hall
ISBN: 978-1-53610-742-5

Chapter 1

RELIABILITY ANALYSIS: THE NEXT STEP TOWARDS RECYCLED AGGREGATES CONCRETE AFFIRMATION

João Nuno Pacheco[1], Jorge de Brito[1,*], Carlos Chastre[2] and Luís Evangelista[3]

[1]CERIS-ICIST, IST, University of Lisbon, Lisbon, Portugal
[2]CERIS-ICIST, FCT, NOVA University of Lisbon, Caparica, Portugal
[3]CERIS-ICIST, TN, IKM, Universitetet i Stavanger, Stavanger, Norway

ABSTRACT

This chapter concerns the need to evaluate the reliability of structural elements produced with recycled aggregates concrete (RAC) in order to address some designers' scepticism towards the use of this eco-friendly material. The current knowledge on RAC's behaviour demonstrates its viability for structural purposes. However, of the investigations performed so far very few are related to a fundamental aspect towards RAC world-wide application as a structural material: structural codes have a probabilistic basis.

After briefly presenting the state-of-the-art knowledge on the material properties and structural behaviour of RAC, the limitations of

* Instituto Superior Técnico, Universidade de Lisboa, Av. Rovisco Pais 1, 1049-001, Lisbon, Portugal. Email: jb@civil.ist.utl.pt.

the current knowledge are debated. Afterwards, an introduction to structural codification is presented, as well as the fundamentals of reliability analysis. Examples of code verifications are contextualized with their underlying assumptions and the information necessary for code calibration is discussed. The role of reliability in the calibration of structural codes is shown, common techniques for reliability calculations are briefly explained, and relevant references in the area are cited for the readers' perusal.

Having established how structural codes are calibrated, the state-of-the-art on the probabilistic and statistical knowledge of RAC properties is reviewed. The implications of the very reduced number of studies on this area are discussed and the need to conduct further studies is emphasized.

Afterwards, investigations that used reliability analysis to calibrate partial safety factors applicable to RAC are reviewed. The methodology of each investigation is presented, the experimental tests that led to the definition of the probabilistic information of the RAC's parameters are described and the need to have a wide range of data coming from different RAC compositions and aggregate sources is debated.

This chapter finishes by contextualizing the current knowledge on RAC properties with the necessary information for code calibration procedures. The relevance of a code proposal towards RAC affirmation as a structural material is highlighted, as well as the requirements of such code. Suggestions for future studies are made.

Keywords: recycled aggregates concrete (RAC), construction and demolition waste, sustainable construction, structural codes; reliability analysis

INTRODUCTION

The sustainability of the construction industry is a matter of pressing concern. Construction activities, particularly concrete production, pose a significant burden on the environment and it is forecast that the environmental impacts of the concrete industry will increase in the coming years, as suggested by predictions on aggregate and cement consumption: the aggregates demand for 2015 was estimated at 48300 million tons and it is predicted that the 2019 demand increases to 51700 million tonnes [1] whilst the United States Geological Survey [2] reported a 2015 cement consumption of 4100 million tonnes, a value that suggests that the 2009 forecast of the World Business Council for Sustainable Development [3] of a 2020 cement consumption in the range of 3700 to 4400 million tonnes may be exceeded.

Table 1. Aggregates demand by the construction sector in 2005-2015

	Yearly demand			Annual growth rate (%)	
Region	2005	2010	2015	2005-2010	2010-2015
Worldwide demand (millions of tonnes)	27300	37400	48300	6.5	5.2
Tonnes per capita	4.2	5.4	6.6	5.2	4.0
North America (millions of tonnes)	3280	3010	3710	-1.7	4.3
Tonnes per capita	10.0	8.7	10.4	-2.6	3.5
Europe (millions of tonnes)	2920	2630	3050	-2.1	3.0
Tonnes per capita	4.0	3.6	4.1	-2.2	2.9
Asia/Pacific (millions of tonnes)	16000	24750	32600	9.1	5.7
Tonnes per capita	4.0	5.9	7.4	7.9	4.6
Other regions (millions of tonnes)	5100	7010	8940	6.6	5.0
Tonnes per capita	3.4	4.3	4.9	4.4	2.9

Source: UEPG [1] and United Nations [10].

Construction and demolition waste (CDW) is a key factor when the environmental toll of construction activities are analysed, and the European Union (EU) has set a minimum target of 70% by weight of CDW re-use/recycling by the year 2020 (European Union Directive 2008/98/CE [4]). The potential impact of this directive on waste reduction is understood after considering that, as reported by Eurostat [5], 33% of the total 2520 million tons of waste produced in the EU in 2015 are CDW.

According to the same Eurostat report, the second group of activities responsible for waste production is quarrying and mining. These activities account for 29% of the waste production and are mostly related to demands from the construction industry. Since the two activities combined generate 62% of all waste within the EU, measures to reduce both activities would have a significant impact towards sustainability.

The aggregates demand by the construction sector, closely related to quarrying and mining activities is shown in Table 1. Worldwide aggregate demand in recent years has been mostly influenced by developing regions, with infrastructure and housing development being the main sources of demand. Nevertheless and despite the recent crisis in the construction sector, the aggregates demand in developed countries is also very significant, with a European *per capita* demand of about 4 tonnes/year.

Table 2. Estimated aggregates production (in millions of tonnes) by source in the EU and in all 39 countries monitored by the UEPG in 2014 [6]

Country	Sand and gravel	Crushed rock	Marine and manufactured aggregates	Recycled aggregates	Re-used on site	Manufactured aggregates	Total production
39 Countries	1434	1985	58	209	19	61	3766
EU 28 only	1004	1169	56	202	19	61	2510

A significant amount of the aggregates used by the construction sector are destined to the concrete industry (55% as reported by the Union Européenne des Producteurs de Granulats -UEPG [6]), since concrete has about 70-80% of aggregates (by volume) in its composition. Thus, the construction sector in general, and the concrete industry in particular, are responsible for a significant consumption of natural resources, mostly due to aggregates' extraction in quarries and riverbeds - as shown in Table 2, about 92% of all aggregates used in the construction sector in the 39 countries monitored by UEPG come from natural resources extraction. The detrimental effects of aggregate extraction are various [7, 8] and range from impacts on wildlife, ecosystem effects (changes in river courses, landscapes and biodiversity) and even structural damages that compromise safety, such as erosion of columns of bridges [9].

A practical way for the concrete industry to minimize its environmental impacts is recycling CDW as recycled aggregates (RA) for the concrete industry: the two major activities of CDW production in the EU would decrease, contributing to the target of 70% of CDW recycling, and the extraction of natural aggregates would decrease. Moreover, studies on the use of CDW on structural concrete have shown that masonry and concrete waste are adequate sources of RA, especially the latter [11]. A report commissioned by the EU [12] explicitly states that from the 520 million tonnes of CDW that are expected to be produced during 2020 inside the EU about 85% will be concrete and masonry, presumably adequate for recycled aggregates concrete (RAC) production.

State-of-the-Art on RAC

The knowledge of RAC properties and behaviour will be analysed in terms of three different themes:

- The material (physical, mechanical and durability) properties and the structural behaviour of RAC;
- The probabilistic/statistical properties of RAC;
- The reliability assessment of structural elements made with RAC.

The first theme will be briefly reviewed, focusing on the differences between RAC and natural aggregates concrete (NAC). The types of experiments performed in RAC so far will be highlighted. The other two themes will be analysed in latter sections of the chapter.

The typical production of RA is similar to how raw stone is crushed into natural aggregates (NA): CDW is crushed, usually in a primary or binary process (using jaw and/or hammer mill crushers), and particles with shape and grading somewhat similar to NA are obtained. Prior to crushing, and depending on the requirements of the particular job, CDW is screened to remove metals and other contaminant materials.

Several different CDW sources have been experimented as RA for concrete: amongst others, Bravo et al. [13] investigated the material properties of RAC produced from unscreened RA from CDW plants; Topçu and Canbaz [14] tested concrete with recycled glass; Valadares et al. [15] and Bravo and de Brito [16] investigated the mechanical and durability behaviour of concrete with tire rubber as partial fine and coarse aggregate replacement; recycled ceramic aggregates have also been tested on concrete [17, 18].

Since the current knowledge on recycled concrete aggregates (RCA) is much more substantial than on all these other sources, RCA will be the focus of this chapter.

RCA are produced from crushing concrete elements, which implies that these RCA have the same physicochemical characteristics of NAC. Their incorporation affects concrete properties since whilst NA are granular materials made from crushed stone, RCA are composite materials made up of fragments of NA and mortar attached to them (Figure 1).

Figure 1. Left: natural aggregates. Right: recycled concrete aggregates.

The main differences between RCA and NA are the much higher water absorption of the RCA [19], the worse mechanical properties - the mechanical strength [20] and the Young's modulus [21] of RCA are lower than those of NA, and the trend for RA to have rougher surfaces and more elongated shapes.

All these particularities of RCA have consequences:

- The water content of concrete with recycled concrete aggregates (RCAC) compositions has to be increased because of the water absorption of the RCA. A distinction between apparent water/cement ratio (w/c ratio) and effective w/c ratio is made, with the latter referring to the water that is not absorbed by the RCA;
- Oftentimes the effective w/c ratio of RCAC compositions has to be higher than that of NAC because of the angularity and roughness of RA;
- This increase in the effective w/c ratio causes a generalized decrease in most concrete properties;
- Furthermore, other properties of the RCA, such as their reduced Young's modulus and increased porosity, lead to a further decrease in some concrete properties, particularly those dependent on aggregate stiffness (Young's modulus [21], shrinkage [22]), and those related to the ingress of aggressive agents [23].

Different researchers have tried diverse beneficiation processes of RCA. These processes follow the reasoning that since the attached mortar is detrimental to RAC properties, if the mortar content of the RCA is decreased, these unfavourable effects are also decreased. Examples are thermal heating [24], microwave heating [25], acid pre-soaking [26], and surface treatment [27]. Another technique that reduces the content of attached mortar and leads to more spherical shapes, even in comparison with NA [28, 29], is the production of RCA with tertiary crushing stages. Since beneficiation techniques can be costly and energy demanding and the use of tertiary crushing is seldom used in research, the knowledge on properties of concrete made with RCA produced with primary and secondary crushing is much more substantial.

Another approach towards reducing the effect of incorporating RCA in concrete is the consideration of the content of attached mortar when designing the concrete mix: in [30] it was shown that if the attached mortar is considered as part of the paste content (leading to a decrease of the volume of new paste), the hard-state properties of concrete are negligibly affected by RCA incorporation at the expense of workability loss.

Since fine RCA typically contain higher fractions of attached mortar, concrete with fine RCA is more affected than RCAC incorporating only coarse RCA, both in durability and mechanical terms [31]. Because of this, there are fewer studies on fine RCAC than on coarse RCAC and most regulations seriously restrict, or bar, the use of fine RA [32].

Since the current state-of-the-art is focused on coarse RCA produced from either primary or secondary crushing without beneficiation, this chapter will deal with this type of aggregates.

By analysing research on the material, mechanical and durability properties of RAC, one can easily conclude that RAC has, in most cases, worse properties than analogue NAC compositions. Since research on the material properties of RAC is readily available [11, 21, 22, 28, 32-54], the state-of-the-art on the matter will not be thoroughly discussed in this document.

The behaviour of structural elements made with RCAC with coarse RCA has also been studied. Most studies are made on structural elements. Li [55] and Xiao et al. [56] reviewed the state-of-the-art of the structural behaviour of RCAC in the People's Republic of China and the suitability of this material was proven. The same conclusions were drawn by other authors in specific studies concerning beams subjected to bending [31, 57] or shear [58], columns [59], and column-beam joints [60] under cyclic loading. The seismic behaviour of scaled bi-dimensional [61] and three-dimensional [62] RCAC frames has been tested, as well of three-dimensional full-scale RCAC frames [63]. The dynamic response of these full scale frames has also been characterized [64] and was similar to the response of a reference NAC frame.

Moreover, it has been suggested that onsite execution conditions are more relevant to the structural response of reinforced RCAC structures than RCA incorporation ratios [65], at least when the RCA are sourced from high quality concrete. However, investigations on curing of RAC under different conditions have shown that the effect of RA incorporation on mechanical [66] and durability properties is marginal [67].

From these studies and structural reasoning, one can conclude that the role of reinforcement steel on the response mechanisms of reinforced concrete significantly reduces the undesirable effects of RCA incorporation: cracking of reinforced concrete members made with RCAC has been experimentally observed as exhibiting similar behaviour to the cracking of reference NAC members [58, 63], and the investigations made so far on the ultimate strength of RCAC elements have shown similar ductility and decreases in load capacity and increases in ultimate displacements below 10% of those reported in reference NAC elements [63].

Most of the studies on the material, durability and structural properties of RCAC suggest that its behaviour is similar to that of NAC. Moreover, it is even argued that structural elements made with RAC comply with structural codes made for NAC. Usually, when analysing the behaviour of structural elements, researchers test a reduced number of specimens of each type of element/material until failure and compare the observed behaviour with code predictions. It is checked whether the code predicted with sufficient accuracy how the material withstood the action imposed during testing. Then, code requirements are compared with the load conditions that lead to different relevant behavioural stages, including failure, and it is verified whether that material complies with the code. One of the most important comparisons made is whether the ultimate load capacity of the experiments satisfies the codified ultimate load capacity.

However, testing whether a structural element made with a new material complies with code formulae is not the same as claiming that code is adequate for structural design using that new material. This apparent paradox is understood if one delves into the underlying concepts of structural codification, the theme of the next section of this chapter. For now, the important aspect is that safety verifications are codified in a way that standardized probabilities of failure are assured.

A relevant aspect of most of the tests made so far on RAC is that, in the case of tests of material properties, usually less than five specimens are tested for each concrete composition and that, in the case of tests on structural elements, only a single structural element of each concrete composition is typically tested. This leads to doubts when assessing the variability of RCAC properties and behaviour that has consequences when inferring the safety of RAC.

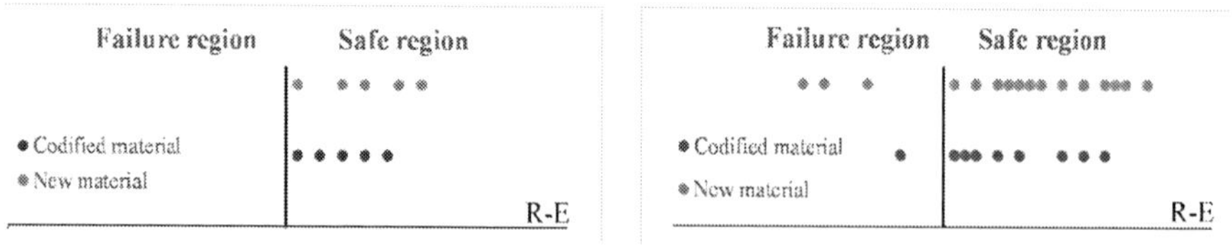

Figure 2. Relevance of number of samples in the assessment of code adequacy.

The evaluation of the compliance of RAC with NAC structural codes using such tests may lead to an overestimation (or an underestimation) of safety: notwithstanding a reduced number of specimens complying with code requirements, if the parameters that control the resistance of RAC structural elements are more variable than the parameters that led to NAC codification, as the number of tests increases, it is expected that more failures of the new material occur. This may be valid even if the expected value of the resistance of the new material is higher in comparison to the material that led to the code definition.

Figure 2 illustrates this concept. This figure concerns the comparison of a hypothetical test on the probability of failure of a structural element made with a new material with the probability of failure of the material for which the code was codified. Specimens that had ultimate limit loads below codified values are shown on the failure region (the R-S axis is the difference between tested ultimate limit load capacity and code predictions). It was intended that the structural element with the new material had the same probability of failure of a similar element made with the material that was already codified.

The hypothetical maximum acceptable probability of this safety verification was 0.04.

As observed in this figure, a reduced number of specimens may lead to erroneous conclusions concerning the applicability of the code:

- When 5 specimens are tested, neither material had samples that failed. Moreover, the new material presented a better overall performance, with a higher average ultimate load;
- When 25 specimens are tested it is found that, notwithstanding the higher average ultimate load of the new material, its number of samples in the failure region was significantly higher than in the case of the codified material. This implies that safety verification is not applicable to the new material.

This phenomenon is caused by the much reduced probabilities of failure of most design verifications. The example in Figure 2 concerns a case with a failure probability of 0.04. In Ultimate Limit-states (ULS) verifications the acceptable probabilities of failure are often lower than 1×10^{-4}, the equivalent to an average of 1 failure per 10000 specimens.

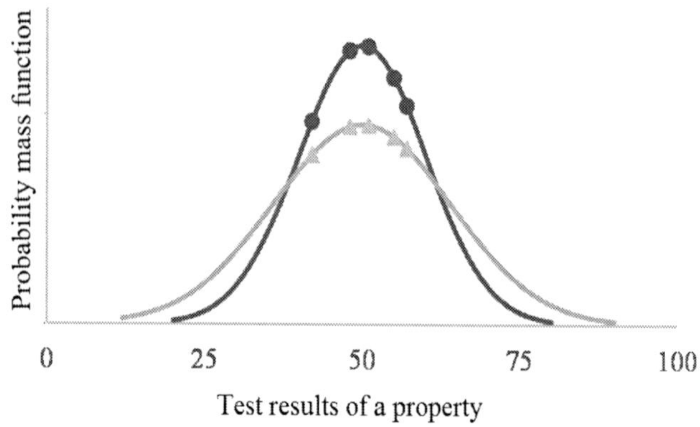

Figure 3. Insufficient knowledge of the lower tail of the probabilistic distribution of a property due to insufficient (five) number of samples.

Since it is completely unreasonable to test such a large number of specimens, tests on material properties are used to make mathematical models of structural response and evaluate safety. These models use the probabilistic distributions of the variable parameters - random variables (RV), involved in the safety problem, and by defining the equation that compares load-effects and resistance, safety is evaluated. This is the theme of the next section of the document.

Nevertheless, the lack of number of tested specimens still hinders code calibration: a significant number of samples is necessary to properly estimate the probabilistic models of the material properties involved in structural verifications. As debated in the next section, the proper estimation of the lower tail of the probabilistic distributions of the material properties is required for ULS codification and, since the probabilistic distributions of material properties tend to have most outcomes concentrated around their expected values, an insufficient number of samples does not give information about the lower (or upper) tail of these distributions (Figure 3).

STRUCTURAL CODIFICATION

Introductory Remarks

Different codes have different criteria and approaches towards uncertainty. However the underlying principles behind their formulation are the same. This chapter does not intend to discuss the calibration of a particular code, focusing on the general principles and criteria used when producing structural codes. Nevertheless, since the discussion of what are mostly theoretical concepts would benefit from illustrative examples of calibration criteria and from how engineering judgment plays a fundamental role in what would otherwise become a purely mathematical process, codification aspects of Eurocode 0 [68], Eurocode 2 [69], ACI 318 [70] and ASCE/SEI 7-10 [71] are presented and discussed whenever considered appropriate.

Code specifications hide an undeniable fact: it is impossible to guarantee that a structure will not fail. The parameters that define load-effects and resistance are uncertain and their randomness leads to cases where, even with correct design and construction, structures fail.

Thus, the role of codes is providing exhaustive sets of specifications that are intended to guarantee that probabilities of non-compliance with performance requirements (probabilities of failure) are within society-accepted target values. Since the direct calculation of these probabilities is often complicated, if not impossible, to compute, design codes mostly rely on deterministic equations, on the specification of minimum material properties, on detailing requirements and on "deemed to satisfy" clauses in order to guarantee adequate reliability without disproportionate costs. The main role of codes is the assurance of standardization amongst practitioners, the improved use of resources, and the presentation of a rational basis to design decisions. Codes transfer responsibility from the designer to the code developers, provided that the code requirements are respected.

The uncertainties in resistance and load-effect definition have different origins that range from randomness in material properties and environmental actions, human action (including construction, utilization, design and maintenance), and modelling.

A better understanding of structural uncertainty is understood after considering uncertainty as a function of three different factors [72]:

- Physical factors are related to the randomness of materials, human actions, environmental conditions, and so forth;
- Statistical factors are those that result from the lack of perfect statistical information - this type of uncertainty increases with a decrease in the number of data (e.g., test samples) available, since it is caused by sparse information;
- Model uncertainty factors result from judging design situations with idealised mathematical models and conditions. For instance: the actual

behaviour of a structural element is different from theoretical response models and standardized test procedures; the constitutive relationships of material properties prescribed in standards follow simplified assumptions; hypothesized load conditions and load history differ from reality.

In Table 3 a set of different causes of the uncertainty introduced in concrete strength modelling are classified by factor. In this table the physical factors were grouped: the ones shown in the first row relate to intrinsic variability of materials - they can never be completely eliminated and their reduction would require the use of more homogenous raw materials; those of the second row concern execution quality related to concrete mixing operations and can be reduced with higher quality control during the corresponding activities; the last group is dependent on quality control during placement and on environmental conditions and human activities that affect curing.

Onsite activities are subject to reduced quality control when compared to activities done in controlled environments. Because of this, the variability in properties of precast concrete is lower than that of ready-mixed concrete, which in turn is a more homogenous material than concrete produced onsite.

Table 3. Classification of different causes of scatter of concrete strength modelling by factor

Physical factors	Statistical factors	Modelling factors
Variability in the material properties of cement, aggregates, admixtures, and others	Estimation of concrete properties using a finite number of samples	The mathematical models of constitutive relationships between concrete properties are not perfect
Differences in concrete composition (between batches and from the prescribed composition) due to weighting, loss of material from scale to mixer, imperfect mixing, …	Impossibility of testing all different concrete compositions	Approximations in load-effect modelling and in theoretical formulations of resistance mechanisms
Different vibration equipment and duration; dependence off the effectiveness of vibration with the concrete composition. Geometric errors when placing formworks and spacers. Differences in curing conditions, including environmental. Reinforcement placement	Constitutive and load-effect modelling are made based on results from a finite number of samples, possibly produced with different concrete compositions than those of the structure being designed	Standardized tests are an imperfect representation of structural concrete conditions

The poorer quality control of onsite activities is also a reason why concrete is a significantly variable material when compared to structural steel: structural steel is produced under very high quality standards, both in terms of material properties and geometrical dimensions.

From the classification shown in Table 3 follows that the uncertainties in RAC strength and structural resistance differ from those of NAC: RA are expected to be more heterogeneous than NAC (physical factor), they have been subject to significant less tests than NAC (statistical factors) and their constitutive models are presumably different than those of NAC (statistical and modelling factors). Other modelling factors might also be different and have not been investigated. For instance, the higher heterogeneity of RAC will probably imply more pronounced size effects, leading to more significant differences between standard tests on specimens and actual onsite strength.

Despite the focus of this chapter being on the effects of the uncertainty concerning the material properties of RCAC on the structural design of reinforced concrete, virtually all other parameters affecting structural design also have a certain degree of uncertainty. For instance, loads have a random nature in magnitude and duration, load distribution models are approximations of reality - e.g., there is no such thing as a uniformly distributed load, and the models that define load-effects have simplified assumptions that do not represent reality accurately, such as idealised support conditions of structural elements. Because of their random nature, most parameters that are involved in design verifications are in fact RVs of uncertain outcome.

The consideration of the possible outcomes of these RVs is implicitly made when structural designers check the different requirements of codes:

- Sets of seemingly deterministic safety verifications that compare specific relevant pairs of load-effects and strength models - partial safety factor formats;
- Deemed to satisfy provisions, such as minimum concrete covers, or restrictions on cement content and cement type based on exposure to external agents.

These requirements are defined by code developers after code calibration procedures that consider relevant design situations that cover a wide range of material properties, actions, environmental conditions, and structural geometries of elements made with the material undergoing codification.

Partial Safety Factor Formats

Safety is checked by limit-state verifications that compare load-effects (E) and structural response (R). A limit-state verification is passed if:

$$E < R \tag{1}$$

If $E > R$ the limit-state verification is not passed and safety is not guaranteed.

There are three categories of limit-states:

- Serviceability limit-states (SLS) - situations that compromise functionality, comfort, and durability;
- Ultimate limit-states (ULS) - partial or total structural collapse is at risk;
- Fatigue limit-states (FLS) - deal with the capacity to withstand repetitive actions.

The three categories of limit-states share the same codification principles. Without loss of generality, this chapter will mostly focus on ULS and FLS will not be discussed.

In the first formats of limit-state verifications, safety was evaluated by verifying whether all critical cross-sections had stresses lower than an allowed stress. The allowed stress was defined as the ratio of the yielding or the ultimate stress of a material by a global safety factor. Thus, the limit-state verification was of the form:

$$E < {}^{R}/_{\gamma_G} \tag{2}$$

where γ_G is the global safety factor.

The underlying principle of Equation (2) is that a deterministic calculation needs some sort of margin to consider variability in loading, worse than expected material properties, and deviations from modelling assumptions and structural response mechanisms. This margin reduces the risk of failure (that is, the probability of failure) if actual conditions differ from the calculation models.

This formulation is not scientifically sound: a global partial safety factor cannot represent the uncertainty in material properties and actions accurately.

Moreover, different design situations using the same global safety factor have very different probabilities of collapse. This happens, for instance, because the use of a single γ_G does not differentiate situations where a structure is subjected to significant variable loads from situations where the dead load would be dominant. Since the uncertainty in magnitude of variable loads is significantly higher, the higher probability of failure in cases of variable load dominance is not reflected by global safety factor verifications. This lack of homogeneity was significantly reduced by using partial safety factors.

Partial safety factors formats differ from global safety factor formats because different safety factors are applied to the different parameters of structural design. The uncertainties in design are accounted for by introducing the concepts of design value of load-effects (E_d) and design value of resistances/structural responses (R_d). Equation 3 becomes:

$$E_d < R_d \tag{3}$$

R_d is obtained by evaluating the response mechanism in terms of design values of material properties and geometry:

- The design value of geometry is usually considered equal to the specified geometry;
- Depending on the code, either material properties or the overall structural response, are reduced in order to consider the uncertainties in the RVs that define the capacity of the structure.

R_d is a function of the factors and variables shown in Equation 4.

$$R_d = R(\eta_i {}^{M_k}/_{\gamma_M}; a_d) \tag{4}$$

where:

- R is the response model of the structure;
- η_i is a conversion factor that accounts for deviations between the material properties in a structure and test results on specimens;
- γ_M is the partial safety factor of the response mechanism. This factor accounts for the uncertainty in material properties (γ_m), in the conversion factor η, and in the structural response model;
- M_k is the characteristic value of the material property;

- a_d is the design value of each geometric variable.

E_d is the design value of the combination of load-effects and is calculated by the combined effect of the multiplication of the characteristic values of actions by partial safety factors and combination coefficients. For instance, the fundamental combination of the Eurocode format defines E_d as:

$$E_d = E\{\gamma_{G,j}G_{k,j};\ \gamma_{Q,1}Q_{k,1};\gamma_{Q,i}\Psi_{0,i}Q_{k,i}\} \tag{5}$$

where:

- E is the load-effect model of the structure;
- $\gamma_{G,j}G_{k,j}$ is the representation of the design value of the permanent loads, modelled as their characteristic values multiplied by respective partial safety factors;
- $\gamma_{Q,1}Q_{k,1}$ is the design action of the basic variable load, modelled as the characteristic value of the load multiplied by the respective partial safety factor;
- $\gamma_{Q,i}\Psi_{0,i}Q_{k,i}$ represents the remaining variable loads, modelled as their characteristic values multiplied by respective partial safety factors and combination coefficients.

The characteristic value of a strength property is a lower bound of its expected properties, whilst the characteristic value of an action is an upper bound.

The choice of the characteristic values is done based on the probabilistic distributions of each RV and on practicability. A characteristic value of actions is typically defined as a value from the 95% or 98% fractiles, whilst characteristic strengths are usually considered as values from the respective 5% fractiles. These values allow the introduction of the uncertainties in material/load characteristics in the design verifications, whilst still being measurable values. Characteristic material properties are conveniently verified by tests and are often specified as guaranteed minimum strengths by producers.

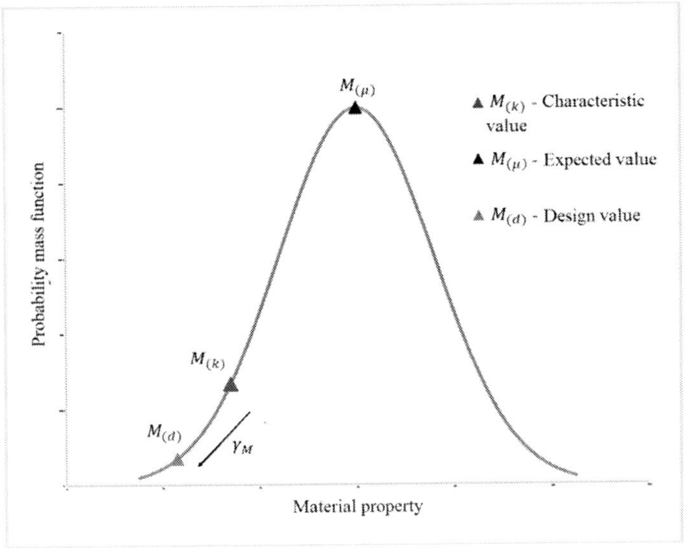

Figure 4. Expected, characteristic, design values of a material property.

Design values are further deviations of the RVs from their expected values (Figures 4 and 5).

Load combination coefficients (Ψ) are:

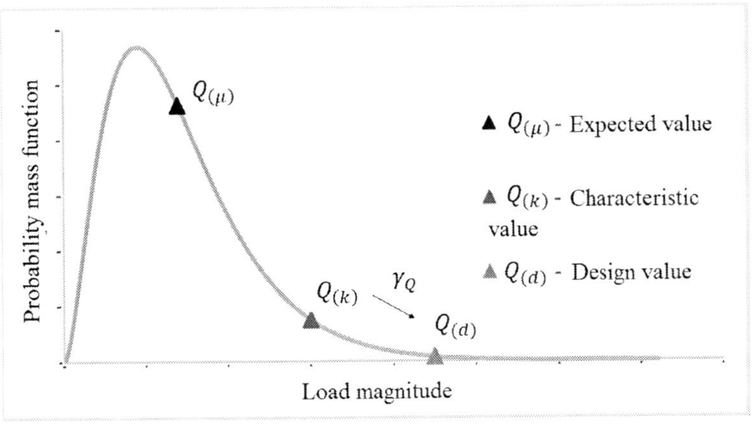

Figure 5. Expected, characteristic, design values of a leading variable action.

- Related to the highly improbable simultaneous occurrence of higher than usual magnitudes of different load components of the structural verification. In the case of ULS, in order to avoid overly conservative designs, the permanent actions and a single leading variable action are multiplied by their partial safety factors. The other variable actions are multiplied by Ψ coefficients that reduce their values (this is equivalent to offsetting the characteristic value of the load components towards the expected value of the actions, reducing the magnitude of the respective design value, as shown in Figure 6);

- Used in SLS, since these design verifications are calibrated for higher acceptable probabilities of failure. In the Eurocode format, Ψ_2 is a combination coefficient applicable to reversible SLS with reduced consequences of failure, Ψ_1 to reversible SLS with some consequences of failure and Ψ_0 to irreversible SLS.

The conceptual generalized representation of Equation (3) is:

$$\beta(G_k \gamma_g) + \lambda_1(Q_{k,1}\gamma_{Q,1}) + \sum_i \lambda_i (Q_{k_i}\gamma_{S_{Q_i}}\psi_{0Q_i}) < \alpha \left({R_k}/{\gamma_M} \right) \tag{6}$$

where:

- The left hand term of Equation 6 concerns load-effect modelling of permanent loads and the following term the load-effect modelling of the leading variable load;
- The following summation represents the combined load effects of the other variable loads;
- The right hand term is the design strength for the case under verification;
- $Q_{k_g}, Q_{k_1}, Q_{k_i}$ and R_k are the parameters that are involved in the verification;
- All these parameters are multiplied by a coefficient (β, λ_i or α) that is a conceptual representation of the theoretical model that quantifies the effect of each parameter on the structural mechanism;
- Load-effects are modelled by multiplying partial safety factors and combination coefficients and strength is divided by partial safety factors;
- Equation 6 is a simplified representation of the format. Resistance mechanisms that depend on more than one source of uncertainty might have different α functions and/or γ_R affecting each R_k.

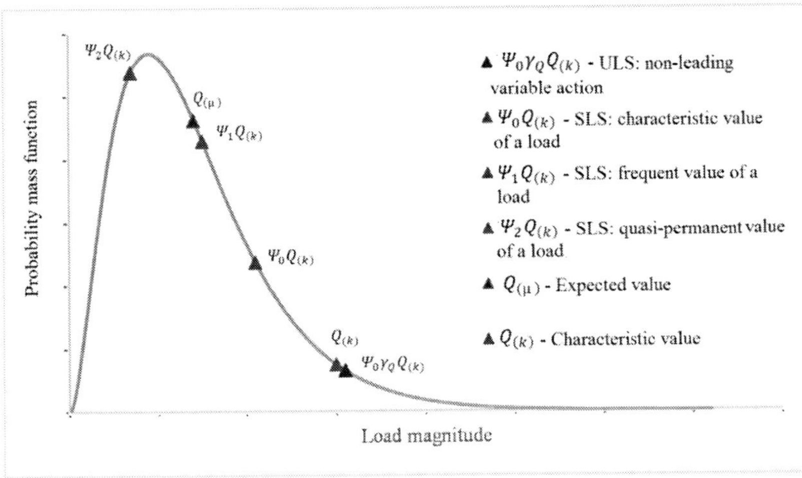

Figure 6. Combination coefficients in ULS and SLS load definition.

This format is based on the probabilistic calibration of the partial safety factors and it is considered that compliance with target reliability indices/probabilities of failure is achieved if the inequality is respected.

The Eurocode format is of the type shown in Equation 6 and a typical ULS verification is represented in Equation 7.

$$\beta(1.35G_k) + \lambda_Q(1.50\,Q_k) < \alpha\left({}^{R_k}/_{\gamma_M}\right) \tag{7}$$

ACI318 [70] and ASCE/SEI 7 [71] share the same principles of the Eurocode formats but some of the safety factors are grouped together in the formulation specified by these standards. All resistance parameters are grouped in the strength reduction factor (Φ) that reduces the load carrying capacity of the member (Equation 8).

$$\Phi R > E_d \tag{8}$$

Equations 9 and 10 concern two of the ULS verifications of ACI318. In the first one emphasis is given to the permanent load; in the latter to the variable loads.

$$E_d = 1.4E_G \tag{9}$$

$$E_d = 1.2E_G + 1.6Q_1 + 0.5Q_2 \tag{10}$$

Table 4 shows the Φ values specified for general reinforced concrete design, exemplifying the dependence of Φ on the failure mechanism.

A fundamental difference between the European approach towards ULS and the North American one is that in the Eurocodes the variabilities are accounted for in the material strength and in the geometric dimensions (the latter are "hidden" in the partial safety factors of material strength). Eurocodes allow the consideration of different factors for the strength of different materials, reflecting differences in variability of, for instance, concrete and steel. On the other hand, the Φ factor of ACI is variable amongst structural member and type of failure (shear, bending, etc.). The Eurocode dependence of γ_M with the material involved in the failure mechanism is that reason why a generalized "γ_M" value was represented in Equation 7. In the case of reinforced concrete, Eurocode 2 [69] uses a γ_C=1.50 for concrete strength and a γ_S=1.15 for reinforcement steel strength.

The definition of E_d shown in Equation (5) is applicable to the fundamental combination of ULS. In the case of SLS, the Eurocodes defines actions as shown in Equation 11.

$$\beta(1.0\ G_k) + \textstyle\sum_i \lambda_{Q_i}(Q_{k_i}\Psi_{k_i}) < {(\alpha R_k)}/{\gamma_M} \tag{11}$$

In Equation 11, Eurocode 0 specifically states that, unless otherwise stated, γ_M should be considered equal to 1.0. Comparing Equations 5 and 10, it is understood that SLS are less stringent than ULS. Partial safety factors do not influence the safety verification and all variable actions are reduced by Ψ coefficients. This is because SLS have lower consequences of failure, thus from a societal and economic perspective, their occurrence is more acceptable, hence their target probabilities of failure are higher than those of ULS.

Partial safety factors were introduced in order to account for uncertainty and ensure that the possibility of non-compliance with performance requirements respects target, society-accepted, probabilities of failure. These target probabilities of failure are defined considering the consequences of failure of each given case. Table 5 shows the ISO2394 - General principles on reliability for structures [73] recommendations for target probabilities of failure and reliability index (β), a concept that will be introduced afterwards.

Table 4. Strength reduction factors of ACI318-14 [70]

Action / structural element	Flexure and/or axial load	Shear and torsion	Bearing	Post-tensioned anchorage zones	Brackets, corbels, struts, ties and nodal zones	Connections of precast members limited by steel yielding	Plain concrete	Anchors in concrete elements
Φ	0.65-0.90	0.75	0.65	0.85	0.75	0.9	0.6	0.45-0.75

Table 5. Target probabilities of failure and reliability indices for different consequences of failure [73]

		Probability of failure (P_f)	Reliability index (β)
ULS	Significant consequences of failure. High cost of safety measure	$1.0x10^{-3}$	3.1
	Significant consequences of failure. Moderate cost of safety measure	$7.0x10^{-5}$	3.8
	Significant consequences of failure. Reduced cost of safety measure	$8.5x10^{-6}$	4.3
Irreversible SLS	Some consequences of failure	$6.7x10^{-2}$	1.5
Reversible SLS	Small consequences of failure	$5.0x10^{-1}$	0

In the initial codifications of partial safety formats, the partial safety factors were often defined arbitrarily on the basis of engineering judgment, previous experience and intuition [68, 74, 75]. This led to cases of different verifications with the same consequences of failure corresponding to significantly different probabilities of collapse, thus different researchers started arguing in favour of the calibration of partial safety factors based on the explicit calculation of probabilities of failure and significant development on the area was made, particularly in the decades of 1970 and 1980 [74-81].

Since partial safety factors have an intrinsic probabilistic meaning, it comes to no surprise that nowadays they are defined based on reliability theory.

Reliability Analysis and Partial Safety Factor Calibration

The calibration of partial safety factors is made by iteratively estimating the probabilities of failure for different sets of design values of the RVs, until the target probabilities of failure are complied with. Afterwards, the partial safety factor of each RV is determined by the ratio of the design value with the characteristic value of the property (Equations 12 and 13).

$$\gamma_Q = {}^{Q_d}/_{Q_k} \tag{12}$$

$$\gamma_c = {}^{M_k}/_{M_d} \tag{13}$$

This process is repeated for a wide range of expectable design situations using that material and type of verification, until a set of partial safety factors that is seen as adequate for a full coverage of the safety verification is defined.

In order to estimate the probabilities of failure of sets of partial safety factors, three types of probabilistic calculation methods are used:

- Full-probabilistic methods - methods that evaluate reliability by the mathematical calculation of the probabilities of failure;
- Reliability index methods - methods that quantify reliability using the reliability index, an alternative measure of safety commonly specified in normative documents and used in code calibration procedures

- Simulation methods - the probabilities of failure or reliability indices are estimated by the generation of random numbers that follow the probabilistic distributions of the RVs involved in the safety calculation. Simulation methods will not be discussed - references [72, 82] explain them.

The use of full-probabilistic methods requires the establishment of a relationship between the different relevant parameters of the design verifications in terms of a limit-state function. This function makes the reliability assessment of one of two situations: failure or safety. Typically g(x) has the format shown in Equation (13).

$$g(x) = Resistance - Load - effects = M \quad (14)$$

Since resistance and action-effects are dependent of RVs, the realization of g(x) is also a RV, named Safety Margin (M). If $M < 0$ the outcome is failure, conversely if $M > 0$ failure does not occur.

When full probabilistic models are used, the probabilities of failure are calculated as shown in Equation (14) and illustrated in Figure 7.

$$P_f = P(M < 0) = \iint_{g\,(x)<0} f_{RE}\,(r, e)dr\,de \quad (15)$$

The calculation of the probabilities of failure is cumbersome and often not possible because of insufficiently precise probabilistic information, lack of closed form solutions, complex computational coding and effort. Another factor that hinders the calculation of probabilities of failure is that for common civil engineering purposes, the target probabilities of failure are extremely small and sometimes within range of numeric integration errors [83].

An alternative to the direct calculation of the probabilities of failure is their estimation in terms of β. This index is commonly used since its calculation is simpler and its principle is intuitive: according to its simplest formulation [76], the reliability index is the distance of the expected value of the Safety Margin to the condition M=0 (the boundary between failure and safety) measured in standard deviations, as presented in Equation (10).

$$\beta = {}^{\mu_M}/_{\sigma_M} \quad (16)$$

where:

- μ_M is the expected value of M;
- σ_M is the standard deviation of M.

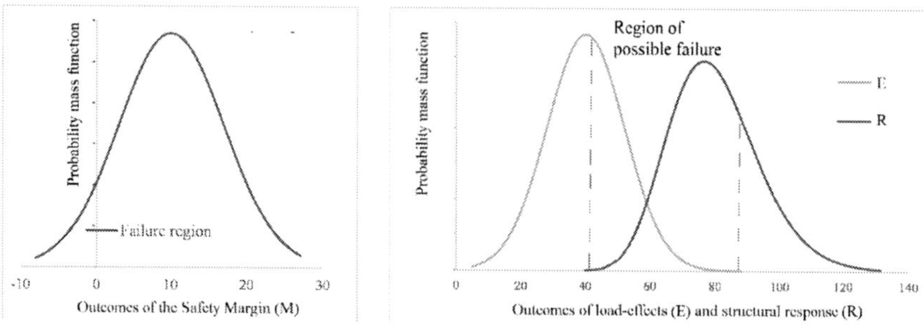

Figure 7. Basic reliability problem: Probability of failure.

Figure 8. Effect of the variability of the safety margin on the probability of failure.

This indirect calculation of the probability of failure considers the expected outcomes of the RVs of the structural problem and also the uncertainties of these RVs. Figure 8 shows two safety margins, M_1 and M_2, of similar reliability problems. Both M_1 and M_2 are normally distributed and have the same expected value but the overall combination of the RVs of M_2 leads to higher uncertainty (hence, standard deviation) than M_1. The higher variance of M_2 increases the relevance of the tails of its probabilistic distribution and, since reliability is conditioned by the distribution about the lower tail, the probability of failure of M_2 is higher, thus the reliability index will be lower, as reflected when using Equation (12).

If the reliability index is defined as proposed by Cornell [76], its calculation only requires the knowledge of the mean values and covariances of the RVs and, if the RVs are normally distributed and uncorrelated, and the limit-state function is of the form of Equation (14), the reliability index is an exact representation of the probability of failure as shown in Equation 17.

$$P_f = \Phi\left(\frac{0-\mu_M}{\sigma_M}\right) = \Phi(-\beta) \tag{17}$$

where:

Φ is the cumulative normal distribution operator.

Since resistances and load-effects are often assumed as uncorrelated, the descriptors of the Safety Margin required for this calculation are:

$$\mu_M = \mu_R - \mu_E \tag{18}$$

$$\sigma_M{}^2 = \sqrt{\sigma_R{}^2 + \sigma_E{}^2} \tag{19}$$

and Equation (16) becomes:

$$\beta = \frac{\mu_M}{\sigma_M} = \frac{\mu_R - \mu_E}{\sqrt{\sigma_R{}^2 + \sigma_E{}^2}} \tag{20}$$

However, in most cases the conditions for the validity of Equation (16) are not met and alternative procedures that allow an estimation of β were developed. First order second moment methods (FOSM) are particularly popular. These are iterative algorithms that use geometry concepts and, in some cases, the probabilistic distributions of the RVs in order to provide an estimate of β. Despite these alternative methods being approximate, they are reasonably accurate for common Civil Engineering purposes [73, 81] and are widely used in code calibration procedures [68, 84].

The explanation of these methods is outside the scope of this chapter. Readers interested in reliability index estimation algorithms are directed to references [78-81, 85-87].

An important aspect of reliability assessment is that it is not intended to estimate the actual probabilities of failure of structures. Diverse factors that are not considered during design, such as construction quality and gross human error, play a significant role in most collapses of modern-day buildings. The calculation of probabilities of failure/reliability indices serves the purpose of providing a notional measure of safety that offers a rational basis towards design processes.

Furthermore, codes are intended to provide a rational basis towards design, providing specifications that comply with the normative minimum reliability levels and are reasonably homogenous amongst different design situations. However, since a code is also expected to be a practical and pragmatic tool, some heterogeneity of β for similar design situations is common.

Probabilistic Models of RAC

Introductory Remarks

It has been argued that structural codification and reliability are intrinsically related. The problem of codification is directly involved with the estimation of notional probabilities of failure that are supposed to comply with target maximum values accepted by society.

As such, the knowledge of the probabilistic information (or in some cases of the statistical descriptors) of the parameters involved in the structural problems is required. In this section, the state-of-the-art knowledge on the variability of RAC properties is reviewed. The analogue information of other properties relevant to structural design is not presented here. In a later section that discusses how a structural code for RAC could be calibrated, they will be briefly referred.

Since the variability in RA characteristics is the reason why it is expected that RAC is more variable than NAC, the first part of this section is a presentation of studies on the composition of RA from different origins (not restricted to concrete waste). The objective of this preliminary analysis is the demonstration that different CDW sources result in RA with different characteristics.

Afterwards, emphasis is given to RCA, since these are the only type of RA that have been sufficiently studied for this purpose. It is shown that even for a single source RA properties vary, namely in what concerns attached mortar content, density, and water absorption.

Having established that RCA is a variable material, all studies known to the authors concerning the probabilistic distribution of the 28-day compressive strength of RAC are presented and benchmarked with the knowledge on NAC properties. The problem of the estimation of properties on structural elements based on data from standardized specimens is discussed at this stage. Finally, the need to research the statistical or probabilistic properties of other RCAC properties is argued for.

The objective of this chapter is to summarize the current state-of-the-art on the probabilistic distributions and statistical descriptors of different properties of RAC, allowing the understanding of expected trends and lacking research. Statistical concepts and probabilistic models are not treated in this document. Readers are encouraged to read specialized publications on the subject, such as [88].

Variability of RA Properties

Any study on the variability of RA sourced from CDW management facilities has to be analyzed considering that regional practice affects the properties of these RA. As Mália et al. [89] has concluded by comparing surveys on CDW composition, CDW is variable by country, a fact that is partly caused by the specific long standing tradition of construction practice of each country.

Nevertheless, in Figure 9 it is shown that the most common sources for RA (ceramics and concrete) account for most CDW across all four countries compared.

The effects of the different compositions of CDW on RA properties are understood after analyzing the studies of Agrela et al. [90] and Bravo et al. [13].

Agrela et al. [90] studied the composition, saturated-surface-dry (SSD) density and 24-hour water absorption of 35 RA produced from 13 CDW plants located on different parts of Spain. These authors divided the RA in three different types:

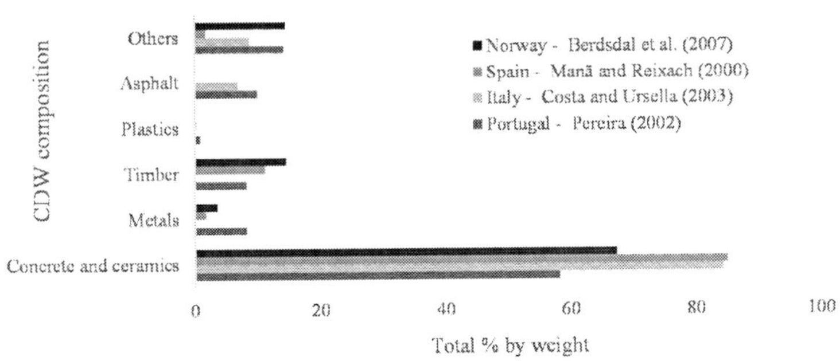

Figure 9. Composition of CDW from 4 different European countries. Data taken from [89].

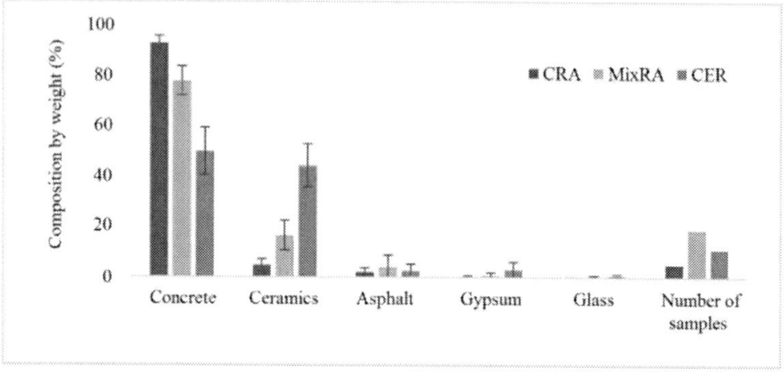

Figure 10. Average and standard deviation of RA compositions sourced from Spanish CDW facilities. Data treated from [90].

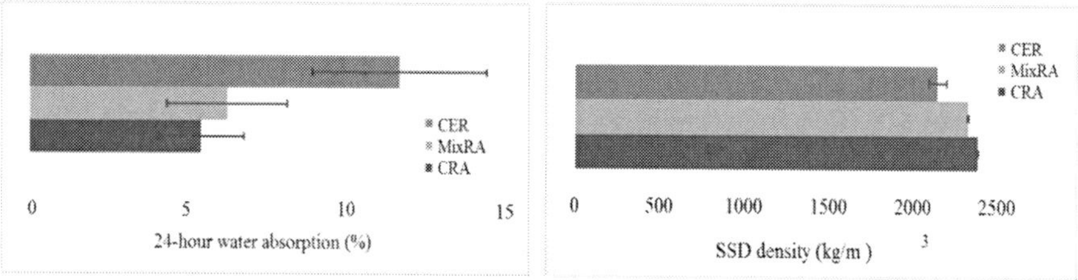

Figure 11. Average and standard deviation of the 24-hour water absorption and the SSD density by aggregate type. Data treated from [90].

- CRA - concrete RA, defined as those that are composed by more than 90% (by mass) of concrete waste (in the remaining of the chapter RCA refers to aggregates sourced exclusively from concrete crushed in laboratory);
- CER - ceramic RA, defined as composed by over 30% (by mass) of ceramic materials;
- MRA - mixed RA, defined as between these two situations.

In Figure 10, a summary of the compositions reported by the authors is presented. The authors reported that CRA have lower water absorptions and higher density than mixed and ceramic RA. In [11] a classification of RA using these two properties concluded that the higher the density and the lower the water absorption of the RA, the better properties RAC has.

Figure 11 concerns the average and standard deviation of the 24-hour water absorption and SSD density. From this figure, two notions are understood: CER have considerably higher water absorption and lower SSD density, and are more variable than CRA. This variability is partly caused by the more variable composition of CER, as shown in Figure 10. MixRA tend to lie somewhat in the middle of the two other types of RA in what concerns average values and standard deviations. The different sample sizes, represented in Figure 10 should be considered when analyzing these data.

Bravo et al. [13] studied the properties of fine and coarse RA sourced from eight Portuguese CDW plants. The CDW was not screened prior to RA production. The average and standard deviations of RA compositions are shown in Figure 12, whilst Figure 13 shows the oven-dry density and the 24-hour water absorption of the RA. The fine RA were sourced from three CDW plants and the coarse RA from five. The authors also made chemical tests and it was concluded that the sulphate content of the RA from one source and the soluble chlorides content from another, made the respective RA unsuitable for concrete.

Figures 11 and 13 show the variability in RA properties from different origins, particularly in what concerns the 24-hour water absorption. This is a key aspect to reason that RA introduce variability in RAC, namely because of the influence of the water absorption of the RA in the local w/c ratio about the ITZs.

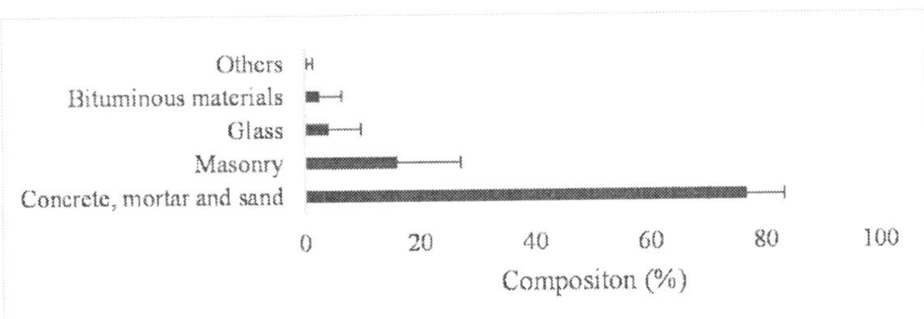

Figure 12. Average and standard deviation of the composition (by weight) of sourced from Portuguese CDW plants [13].

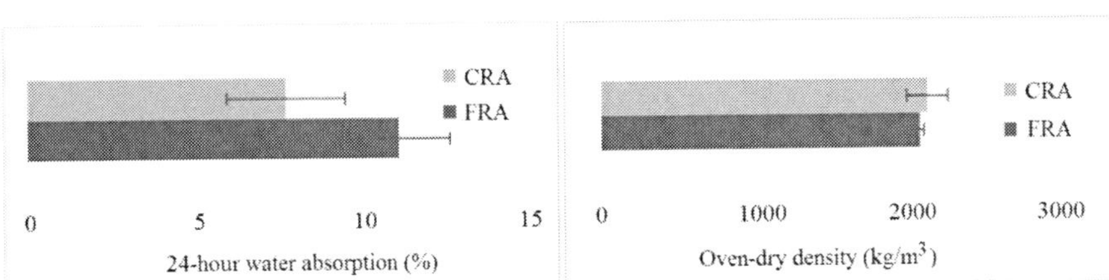

Figure 13. Average and standard deviation of the 24 h water absorption (left) and oven-dry density (right) of RA sourced from Portuguese CDW plants [4].

Having shown that RA are significantly variable materials, studies relative to the properties of RCA will be analyzed:

- RCA have been studied more comprehensively than RA produced from other sources;
- RCA are expected to be the best source of RA for structural concrete, as shown in diverse studies on the expected performance of RCAC and as also suggested in Figure 11: from the three sources, higher RA water absorption (detrimental to concrete properties) and higher variability of RA properties (detrimental to the codified design values of the properties of RAC) are associated with lower concrete content in the RA.

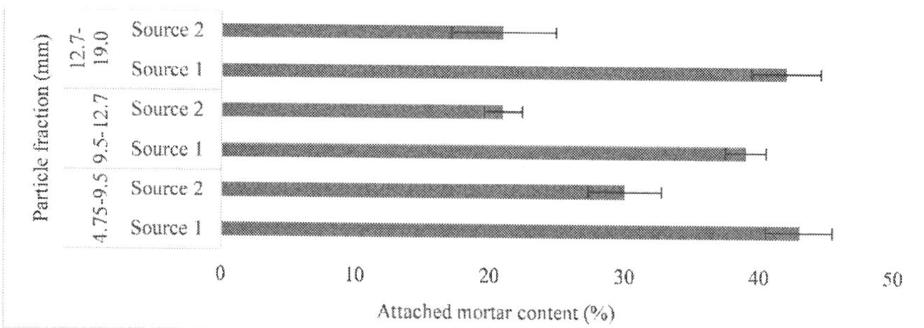

Figure 14. Average and standard deviation of the attached mortar content of RCA sourced from Canadian concrete waste [91].

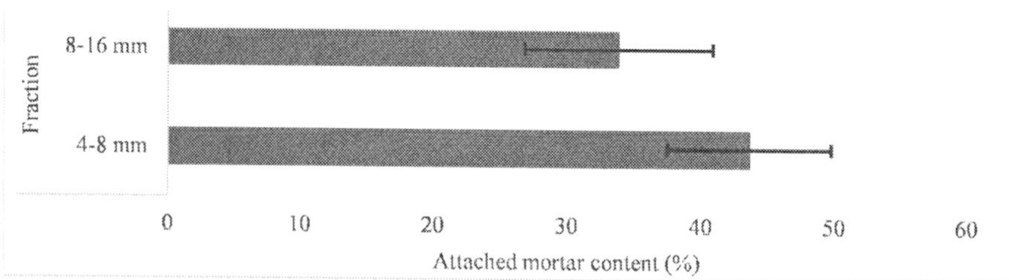

Figure 15. Average and standard deviation of the attached mortar content of the RCA studied in [92].

As discussed in the presentation of RAC properties, the main cause for RCA effects on concrete is the attached mortar: this material is responsible for two additional interfacial transition zones (ITZs), higher water absorption and roughness, and lower aggregate strength. Because of this, studies that compared the variability of the attached mortar content of RCA with RCA properties will be presented herein. This mortar content is not only dependent on the source concrete properties and crushing process, but also on particle size, as shown in Figures 14 and 15 [91, 92].

Butler [93] and Henry et al. [94] compared the average values of SSD density and 24-hour water absorption of NA and of three RCA sourced from concrete with different compositions. As shown in Figure 16, the variability in the expected value of the water absorption across RCA sources is perceived. Butler [95] surveyed different investigations on the subject and reported that the 24-hour water absorption of coarse RCA ranges from 0.57% to 9.25%. In Pepe et al. [96], the 24-hour water absorption, measured in studies conducted solely by these authors, ranged from 4.94% to 11.94%.

Because of the higher heterogeneity of the attached mortar in comparison with NA, the presence of an additional ITZ, and of the variability in attached mortar ratios, the compressive strength of RCA is also expected to be more variable than the compressive strength of NA. Pepe [20] studied this variability and confirmed this hypothesis. Nevertheless, it is well known that aggregate strength only limits the strength of concrete compositions within the high strength range.

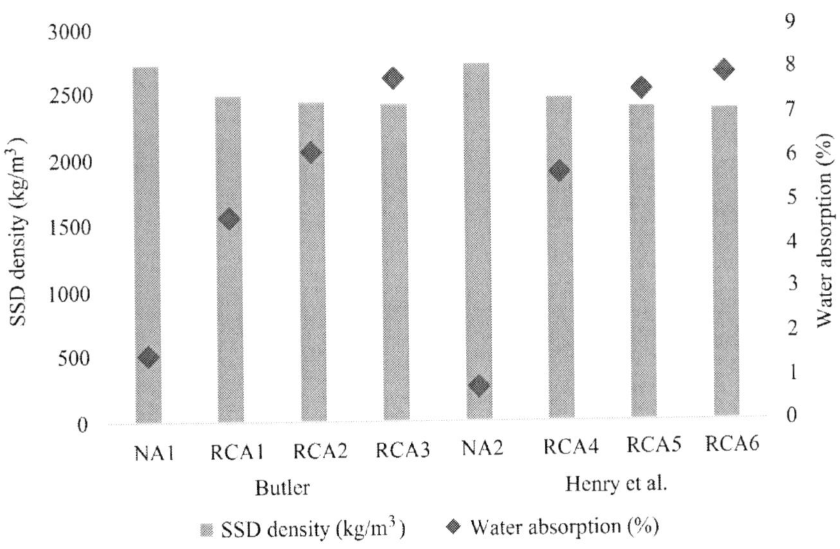

Figure 16. Average values of water absorption and SSD density of the RCA studied in [93] and [94].

Statistical and Probabilistic Studies on RAC Properties

Very few studies on the statistical or probabilistic properties of the 28 day-compressive strength of RAC have been made.

In Table 6, the first and second moment statistical descriptors of studies with over 20 samples tested for each composition are summarized.

When analyzing these results, some considerations should be made:

- Since the specimens are produced in controlled environments, the expected value and CoV of the properties of actual RAC structures are expected to be higher. This is a well-known aspect of concrete quality control [100, 101] that did not hinder the use of analogue data of NAC in the calibration of NAC codes [84, 102];
- No effort to combine the data of different tests into pooled expected values and standard deviations was made since such attempt would be artificially influenced by the concrete compositions and strength ranges chosen by the authors that made the studies surveyed;
- Specific aspects of each of these studies will be discussed next, as well as probabilistic data, whenever available from the studies;
- The results shown in Table 6 concern the production of RCAC using a single source of RCA. RCAC produced using more than one source of RCA will presumably have higher variability, as experimentally reported in [94];
- All these studies concern the incorporation of coarse RCA only.

Xiao et al. [97] studied the probabilistic distribution of the 28-day compressive strength of RCAC. Four different concrete compositions were investigated:

- A reference mix (NAC) - 75 standard cubes;
- A RCAC mix with 30% of coarse RCA incorporation - 100 standard cubes;
- A RCAC mix with 50% of coarse RCA incorporation - 98 standard cubes;
- A RCAC mix with the total coarse fraction replaced by RCA - 98 standard cubes.

The standard cubes had edges of 100 mm and the RCA were produced from a single source of concrete waste. The concrete mixes had a target compressive strength of 30 MPa.

The authors reported that the average compressive strength decreased with RCA incorporation and that the coefficient of variation (CoV) was unaffected. The results were fitted to normal and log-normal distributions and Chi-square tests (significance level of 5%) confirmed that both distributions were suitable. In Table 6 the statistical descriptors of all tests are shown and in Figures 17 and 18 the histograms and normal distributions of two compositions (RAC100 and NAC) are shown.

All mixes had coefficients of variation in the range of 8.2%-9.7%, thus any dependence of the variability on RCA incorporation seems marginal. Nevertheless, there was a trend towards higher variability when both NA and RCA were mixed. Henry et al. [94] reported the same findings.

Table 6. Statistical descriptors of studies of the 28-day compressive strength of RAC

Investigation	Number of samples	RCA incorporation (%)	Average (MPa)	Standard deviation (MPa)	CoV (%)
China - Xiao et al. [97]	75	0	41.6	3.4	8.3
	100	30	41.5	3.9	9.5
	98	50	40.2	3.9	9.7
	98	100	36.5	3.0	8.2
Spain - Etxeberria et al. [98]	24	0	44	3.9	8.8
	14	25	43	4.0	9.3
	43	25	41.5	4.6	11.0
	20	50	46	4.3	9.3
	14	100	46.5	5.9	12.6
	33	100	40	6.7	16.7
China -Xiao et al. [99]	83	100	18.48	2.6	13.8
	83	100	27.16	4.2	15.5
	83	100	43.65	8.0	18.3
	40	100	20.17	1.6	7.7
	40	100	31.09	1.8	5.8
	40	100	46.66	3.2	6.9
Japan - Henry et al. [94]	30	0	41.9	1.23	2.9
	30	50	31.3	1.4	3.9
	30	100	34.4	1.3	3.9
	30	100	32.7	1.3	4.3
	30	100	28.8	1.2	4.4

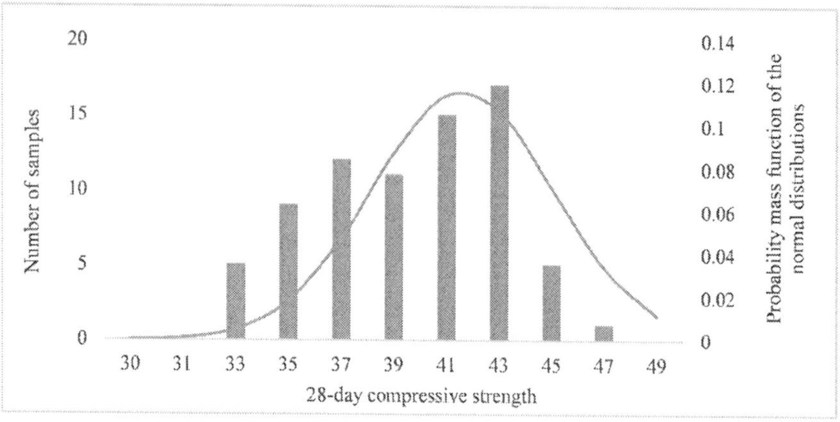

Figure 17. Histogram and normal distribution of the NAC composition of [97].

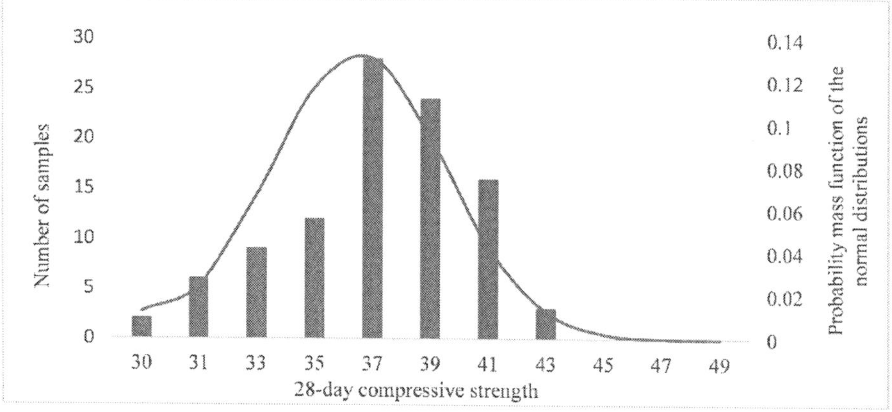

Figure 18. Histogram and normal distribution of the RAC100 composition of [97].

Henry et al. [94] studied the statistical descriptors of the 28-day compressive strength of seven concrete compositions made resorting to four different coarse aggregates - one NA and three RCA from different sources. The mixes produced were:

- NAC - a mix using exclusively NA;
- RAC1 - a mix with total coarse replacement with RCA from source 1;
- RAC2 - a mix with total coarse replacement with RCA from source 2;
- RAC3 - a mix with total coarse replacement with RCA from source 3;

- NAC-RAC1 - a mix with a 50% replacement of coarse NA with RCA from source 1;
- RAC1-RAC3 - a mix with half the coarse aggregates sourced from RCA source 1 and the other half from RCA source 3.

In Table 7 the second-moment statistical descriptors of each mix are shown. The use of RA caused increases in the CoV. Moreover, when different sources were mixed, the CoV increased further, even when one of the aggregates was NA.

The same conclusions regarding mixing RCA sources were reached by Breccolotti and Materazzi [100] that tested a total of 18 compositions using four different RCA and a NA source (10 specimens per test). These authors argue that, even though during mixing the ratio between recycled and natural aggregates is known, the actual ratio poured into the moulds is a RV. This uncertainty in the actual ratio may be a reason for the increased CoV of mixes with intermediate incorporation ratios, since different concrete samples of the same theoretical RA incorporation ratios will, in fact, differ in the actual incorporations of RA. Moreover, even if the RA ratios of two samples were exactly the same, the RA incorporation ratio on the failure surfaces of these samples are not necessarily the same.

In disagreement with the findings of Xiao et al. [97], Henry et al. [94], and Breccolotti and Materazzi [100], Etxeberria et al. [98] reported higher CoVs for full replacement ratios. Different concrete compositions were made: a NAC mix, a mix with 25% of RCA incorporation (RAC25), a mix with 50% of RCA incorporation (R50) and a mix with total RCA incorporation (RAC100). The compressive strength of all mixes was fixed by changing the cement content (hence, the w/c ratio).

Table 7. Statistical descriptors of the 28-day compressive strength tests of [94]

Composition	Average	Standard deviation	CoV
NAC	41.9	1.23	2.9%
RAC1	34.4	1.26	3.9%
RAC2	32.7	1.28	3.9%
RAC3	28.8	1.24	4.3%
NAC-RAC1	31.3	1.38	4.4%
RAC1-RAC3	28.7	1.72	6.0%

Table 8. Statistical descriptors of the 28-day compressive strength tests presented in [98]

	Sample size	Average	Standard deviation	CoV
NAC	24	44.0 MPa	3.9	8.8%
RAC25	15	43.0 MPa	4.0	9.3%
RAC25-2*	43	41.5 MPa	4.6	11.0%
RAC50	20	46.0 MPa	4.3	9.3%
RAC100	15	46.5 MPa	5.8	12.6%
RAC100-2*	31	40.0 MPa	6.7	16.7%

* These compositions were batched twice.

Table 9. Coefficient of variation (%) of the compressive strength of concrete compositions with over 35 MPa of strength as specified in [106]

		Control standard		
		Excellent	Good	Poor
Overall variation	General construction	<7.0	9.0-11.0	>14.0
	Laboratory specimens	<3.5	4.5-5.0	>7.0
Within batch variation	General construction	<3.0	4.0-5.0	>6.0
	Laboratory specimens	<2.0	3.0-4.0	>5.0

The 28-day compressive strength was assessed by testing 20 150 mm cubic specimens and it was concluded that the standard deviation and the CoV increased with increasing RCA ratios. Table 8 summarizes the results of this experiment - the different sample sizes should be considered when analyzing the data.

Comparing the second moment descriptors of these investigations with the assumptions behind standards, a lower CoV/standard deviation of RAC in comparison with NAC is concluded:

- Eurocode 2 [69], as well as Model Code [103, 104], considers the compressive strength of concrete as normally distributed with a standard deviation of 4.86 MPa independent of the strength class. This value is considered after a research made by Rusch [105];
- The Chinese codes JGJ 55 (2011) and GB 50010 (2010) consider the standard deviation of the 28-day compressive strength as 5 MPa [99];
- ACI214 [106] models the compressive strength as normally distributed until strengths of 70 MPa. This document claims the CoV is less affected by strength class than the standard deviation. In this document the differences of standard deviation and CoV of laboratory

produced specimens and on site concrete are quantified. Table 9 shows the standardized CoV values of this document for excellent, good and poor control of concrete compositions with over 35 MPa of compressive strength.

Table 9 justifies the reduced variability of RAC in comparison to NAC: specimens produced under laboratory conditions tend to have lower variability. Moreover, usually laboratory trials only evaluate the within batch variability. In order to properly consider the variability of RAC properties on actual buildings, it is necessary to use models that transpose laboratory tests to onsite concrete, as done in NAC codification [102, 107].

To the authors' best knowledge, the only other RAC property that has been studied in statistical terms is the 28-day splitting tensile strength.

In Xiao et al. [108] the results of an experimental campaign on the splitting tensile strength of RCAC are reported. Figure 19 presents a histogram of the results of this experiment for three different mixes with total coarse RCA incorporation and three different w/c ratios. The tests were made on 83 samples per mix. The authors reported that a normal distribution suitably fitted the experimental data. The statistical descriptors of these tests were reported in [99] and are shown in Table 10.

These results show much higher second moment descriptors in comparison to the variability of the 28-day compressive strength (Table 6). This phenomenon is common to NAC [109].

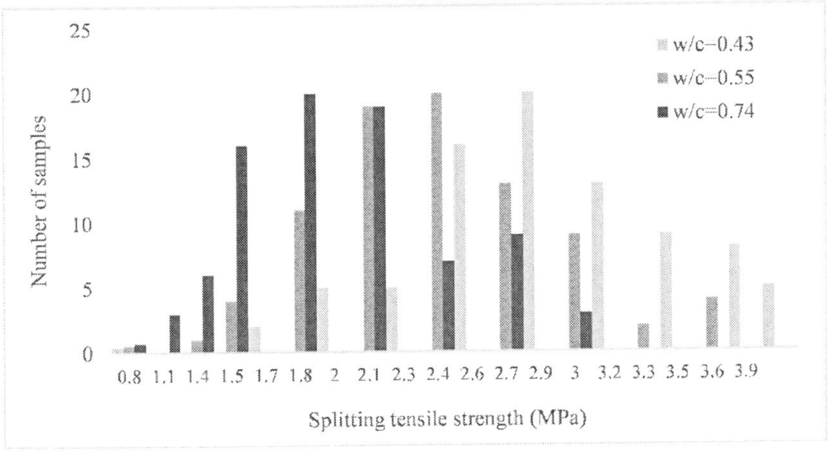

Figure 19. Histogram of the splitting tensile strength tests presented in [108].

Table 10. 28-day splitting tensile strength tests surveyed in [99]

Specimens	RA %	Average (MPa)	Standard deviation (MPa)	CoV (%)
83	100	1.39	0.41	29.3
83	100	1.97	0.09	4.7
83	100	2.65	0.59	22.2
40	100	1.90	0.30	16.0
40	100	2.31	0.33	14.5
40	100	2.83	0.59	20.9

Final Remarks

Additional data on the statistical and/or probabilistic properties of RAC is lacking [99, 100], especially in what concerns the knowledge of the probabilistic distribution of RAC properties. This hinders a proper calibration of ULS partial safety factors, since for ULS the lower tail of the probabilistic models of strength parameters is of the upmost importance [110].

As argued in [75], usually the calculation of the probabilities of failure is more dependent on the choice of the distributions types of the RVs than on the estimation errors of the parameters of the distributions. Physical considerations and an overall judgment of the situation are common and accepted. The central limit theorem argues in favour of modelling material strength of concrete structures following normal distributions since common concrete structural verifications involve ductile response mechanisms, leading to an overall additive behaviour.

The 28-day compressive strength of RAC has been studied to some extent but no comprehensive experimental campaign covering different compositions has been made and the very different number of samples across investigations interferes with a proper comparison between them.

The other concrete properties have not been properly studied using a statistically relevant number of samples and it would be important to relate the probabilistic knowledge of the compressive strength of RAC compositions with the respective statistical data of other RAC properties of the same composition, perhaps in a study that would also investigate the constitutive relationships of the most relevant RAC properties.

Since the variability in attached mortar of the RCA is probably the major cause for higher scatter in RAC properties, concrete properties that are largely dependent on aggregate stiffness, such as creep, shrinkage and Young's modulus are expected to present significant variability. Coincidently, these properties are amongst the concrete properties that are the most affected by RCA incorporation on average, thus the combined effect of worse first and second moment descriptors may be especially relevant for the structural design of RAC.

State-of-the-Art on the Reliability of RAC Elements

The reliability evaluation of RAC structural elements is somewhat compromised by the limitations concerning the probabilistic and statistical knowledge discussed on the last section.

Given this lack of knowledge, little research has been made on this theme and most authors that investigated this area opted to follow assumptions under worst case scenario hypothesis, rather than data from actual experimental tests.

Breccolotti and Materazzi [111] made analytical calculations concerning the reliability of the bond strength between RAC and structural steel using the bond slip relation proposed on Model Code 2010 [104]:

$$\tau_m = 2.5\sqrt{f_{ck}} \tag{21}$$

The criteria that defined the limit-state function was that failure occurred if the bond strength was lower than the design value of the yield strength of the reinforcement:

$$g(x) = R - E = \int_0^l \pi\phi\ \tau(f_c, x)dx - A_s f_y \tag{22}$$

where:

- l is the anchorage length;
- f_c is the RV of the concrete's compressive strength;
- The integral is made across the anchorage development.

The partial safety factors of concrete strength were calibrated with the objective of determining the anchorage length of RAC that ensures the same reliability level of NAC. This target reliability was fixed at 3.5. The CoV of the compressive strength of RAC was arbitrarily considered as varying from

15% to 30% in order to account for the presumed higher variability of RAC in comparison to NAC.

The authors concluded that the partial safety factors obtained (γc in the range of 1.46-1.57) do not justify the adoption of a partial safety factor different than used in NAC ($\gamma_c = 1.50$).

Woerner and Moerland [112] investigated the reliability of RAC beams designed according to Eurocode 2 [69] subjected to bending. The methodology of this investigation was:

- The steel yield strength and concrete compressive strength were modelled as lognormally distributed, whilst live and dead loads were normally distributed. The cross-section geometry was considered deterministic;
- The CoV of a NAC composition (strength class C20/25) was estimated using the mean compressive strength and the 5% fractile of this property specified in Eurocode 2 [69];
- Arbitrary changes of the statistical parameters of the compressive strength of the C20/25 mix, simulating two RAC mixes. Calculation of the corresponding CoV (maximum CoV of 26%);
- A common cross-section geometry of beams was defined, as well as different reinforcement ratios (between 0.1% and 3.0%). The failure mechanism (by tensile steel strength or compressive concrete strength) was dependent on this ratio;
- The limit-state function was defined as the difference between the ultimate moment-carrying capacity and the acting bending moment;
  $$g(x) = M_R - M_E \qquad (23)$$
- A reliability analysis using first order reliability methods was conducted and the partial safety factor γ_c was calibrated for the different conditions of loading, reinforcement and concrete variability;
- The authors reported that the partial safety factor γ_c prescribed in Eurocode 2 (1.50) produced β values always above Eurocode 0 requirements.

Xiao et al. [99] studied the amount of reinforcement that had to be added to beams subjected to bending in order to achieve similar β values to those of NAC beams. From this investigation, the following aspects should be highlighted:

- After surveying the state-of-the-art on the statistic
- al properties of the 28-day compressive strength, the authors made the reliability analysis considering that the CoV of the 28-day compressive strength was 20%, using pessimistic expectations;
- β was calibrated after the Chinese Code GB50153-2008;
- Concrete and reinforcement strength were modelled as lognormally distributed;
- Dead and live loads were respectively modelled as normally distributed and extreme type I distributed;
- As in the case of [112], the limit-state function was defined as the difference between the ultimate moment-carrying capacity and the acting moment;
- As expected, the authors reporter that β values decreased with the increase of the CoV of the compressive strength of RAC, especially for beams with reduced reinforcement ratios (ultimate cross-sectional load-carrying capacity conditioned by concrete compression failure);
- The maximum increase of reinforcement steel required to ensure the same reliability indices was 3.5%.

Breccolotti and Materazzi [100] calibrated the partial safety factor of concrete strength of RAC columns. The methodology of their calibration process was the following:

- The authors produced four groups of concrete mixes with three different sources of RCA and evaluated the respective statistical parameters of strength (mean, standard deviation and CoV). These results were not shown in Table 6 because the number of specimens tested was not disclosed;
- It was reported that the incorporation of RCA resulted in higher scatter of compressive strength, except when very good quality RA were used. In that case, the CoV was reduced by incorporating RCA. The maximum CoV of the RCA mixes was 4.5%.
- A typical cross-section of a column was designed;
- A reliability analysis of the NAC mixes was made using the method proposed by [78]. Three RVs were considered as either lognormally distributed (steel yield strength and concrete compressive strength) or normally distributed (the axial load). The means and standard deviations of these parameters were defined considering previous

probabilistic studies on common loads and resistance models, except the concrete compressive strength, whose parameters where defined after the experimental results. The calculation of β considered three cases: zero eccentricity, moderate eccentricity and high eccentricity;

- The β values of the NAC columns (for $\gamma_c = 1.5$) were higher than specified in Eurocode 0. The authors claim this was due to the small CoV values of the experimental results;
- An iterative process where the RAC partial safety factors were calibrated was made. The calibration criterion was the reliability index of the RAC columns being equal to the β calculated for the NAC columns;
- The increases in the partial safety factor of RAC were directly related to the quality of the RA and were more relevant when the concrete was under pure compression (maximum γ_c of RAC of 2.1), since when columns are subject to bending the reinforcement steel has a relevant role in structural behaviour (γ_c of 1.65 when the load had maximum eccentricity, the incorporation ratio was total and the analysis was made with data from the RAC mix with the highest scatter).

Table 11. Summary of the studies on the reliability of RAC

Authors	Type of verification	Objective	Max. CoV σ_{cRAC}	Basis for CoV σ_{cRAC}	Conclusion
Breccolotti and Materazzi [111]	ULS bond	γ_c for anchorage length. $\beta_{RAC}=\beta_{NAC}$	30%	Assumed as worse than CoV σ_{cNAC}	Max γ_c=1.57
Woerner and Moerland [112]	ULS beam flexure	γ_c calibration for RAC. $\beta_{RAC}=\beta_{NAC}$	26%	Assumed as worse than CoV σ_{cNAC}	γ_{cNAC} adequate for RAC
Xiao et al. [99]	ULS beam flexure	Change in reinforcement ratio for $\beta_{RAC}=\beta_{NAC}$	20%	Higher CoV surveyed	Maximum reinforcement increase of 3.5%
Breccolotti and Materazzi [100]	ULS columns	γ_c calibration for RAC. $\beta_{RAC}=\beta_{NAC}$. Different eccentricities	4.5%	Laboratory tests (β_{NAC} calculated identically)	γ_c = 1.65 - high eccentricity γ_c = 2.10 - pure compression

In Table 11, the current limitations on the state-of-the-art reliability of RAC are understood:

- Very few types of verifications have been investigated and all address ULS. Moreover, the approaches of these investigations differ and their results are not consensual;
- Only the effects of the variability of the compressive strength have been investigated;
- All reliability analysis made so far are hindered by the lack of knowledge on the statistical and probabilistic information concerning the RAC properties.

CONCLUSION

The use of RA in structural concrete is a step towards a more sustainable concrete industry. This solution enables a reduction of CDW and natural aggregate's consumption. The knowledge on RAC properties is already substantial, especially in what concerns RCAC. Nevertheless, designers, contractors and clients are unsure about the use of this material, since RCAC is looked upon as a material with uncertain properties.

In fact, notwithstanding the different studies on the material, durability and structural properties of RAC made so far, research on the probabilistic and statistical models of RAC properties is insufficient. The research on RAC has been focused on the expected performance and very few studies on the variability of RAC properties have been made, which partly explains the doubts of the construction sector.

Moreover, few reliability analyses on RAC structural behaviour have been made, and most of these analyses were made with underlying assumptions concerning the variability of RAC material properties that have not been scientifically justified. Since a key aspect of structural codification is the calibration of structural materials towards standardized reliability and safety, full-scope studies in this area would be a significant step towards RAC acceptance by designers, constructors and clients, leading to a development of a code that would provide the necessary work frame for structural design using RAC.

An industry-accepted code for RAC design will need to be simple and:

- Be supported by robust and representative data;
- Cover a wide range of design situations;
- Follow the same assumptions and criteria behind NAC codes;
- Be based on the format of an already existing concrete structural code with as few changes to its sets of verifications and provisions as possible.

The present-day state-of-the-art on RCA incorporation on structural concrete has already established the physical and chemical mechanisms that are responsible for the changes in concrete properties, as well as the expected values of most properties of RCAC necessary for common structural design. However, two distinct types of codified information are lacking: the constitutive modelling of RCAC properties and the assurance of proper reliability.

Constitutive modelling was not treated in this chapter. Different approaches have been made by different authors: some models are accurate but were defined for specific concrete compositions, others are based on meta-analysis of several investigations and are simple, intending to cover a wide range of concrete compositions [21, 22, 34-38], and other approaches attempt to use several investigations and use complex models in order to provide more accurate results [113].

Since codes should be easy to implement and avoid situations prone to users' errors, so far codes have focused on simplification: the constitutive modelling of NAC is also made resorting to simple equations and few parameters (most models that estimate concrete properties are only dependent on the compressive strength) and significant scatter between the actual properties and the properties estimated by the models is accepted [109].

The errors introduced in the safety verifications because of constitutive modelling should be accounted for when defining model uncertainties (Table 3). The scatter of the constitutive models of RAC and the experimental properties will have to be analysed and compared with the analogue scatter introduced by the models of NAC. This deviation between actual and assumed properties should be modelled as a RV affecting resistance models.

The assurance of proper reliability requires more knowledge on the probabilistic and statistical properties of RAC. Investigations that relate different concrete properties using a significant number of samples are lacking and would be an important contribution towards the evaluation of RAC reliability.

This is especially relevant given the fact that the reliability of SLS and of the durability of RAC have not been studied and some of the concrete properties that are most affected by RA incorporation are involved in such verifications.

In what concerns other factors that could affect the reliability of RAC, perhaps it is too ambitious to expect studies on the specificities such as the effects of curing conditions and differences in specimen size on RAC properties at this stage. Despite the more heterogeneous nature of RA probably leading to more pronounced size effects on RAC than on NAC, it seems reasonable to follow the assumptions used in NAC modelling when transposing laboratory tests of samples to onsite material properties.

Geometrical deviations, actions, reinforcement strength and model uncertainties caused by load-effect modelling and resistance models can also be accounted for using their analogue NAC models. Several different studies on the probabilistic distributions of these parameters have already been made and used for NAC calibration [72, 74, 101, 114-121].

REFERENCES

[1] Freedonia, 2012, World construction aggregates, in Industry. Study No. 2838, The Freedonia Group, Ohio, USA, 334 pages.

[2] United States Geological Survey, 2016, Mineral commodity summaries - Cement, Virginia, United States of America, 2 pages.

[3] CSI, 2009, The Cement Sustainability Initiative: Recycling concrete, Geneva, Switzerland, 42 pages.

[4] EU DIRECTIVE 2008/98/EC on waste, 2008, European Union, Brussels, Belgium, 28 pages.

[5] Eurostat. Waste statistics in Europe. 2015 Accessed on October 2015 via the website: http://epp.eurostat.ec.europa.eu.

[6] UEPG - Union Européenne des Producteurs de Granulats, 2014, Estimates of aggregates production data. Accessed on October 2015: http://www.uepg.eu/ statistics/estimates-of-production-data/data-2014.

[7] Basavarajappa, H. T., M. C. Manjunatha, L. Jeevan, 2014. Sand mining, management and its environmental impact in Cauvery and Kabini river basins of Mysore District, Karnaka, India using geomatics techniques. *Int. J. Civ. Eng. Technol.* 5(9), 169-180.

[8] British Geological Survey Commissioned Report CR/03/029N, 2003. River mining: aggregate production and supply in developing countries. University of Exeter, 96 pages.

[9] Sousa, J. J., L. Bastos, 2013, Multi-temporal SAR interferometry reveals acceleration of bridge sinking before collapse. *Nat. Hazards Earth Syst. Sci.* 13(3), 659-667.

[10] United Nations, 2015, World Population Prospects: The 2015 Revision, Department of Economic and Social Affairs. Custom data acquired on October 2015 via the website: https://esa.un.org/unpd/wpp/DataQuery/

[11] Silva, R. V., J. de Brito, R. K. Dhir, 2014. Properties and composition of recycled aggregates from construction and demolition waste suitable for concrete production. *Constr. Build. Mater.* 65, 201-217.

[12] BIO Intelligence Service, 2011. Service contract on management of construction and demolition waste - SR1: Final report task 2, European Commission. 2011.

[13] Bravo, M., J. de Brito, J. Pontes, L. Evangelista, 2015. Mechanical performance of concrete made with aggregates from construction and demolition waste recycling plants. *J. Clean. Prod.* 99, 59-74.

[14] Topçu, I. B., M. Canbaz, 2004. Properties of concrete containing waste glass. *Cement Concrete Res.* 34(2), 267-274.

[15] Valadares, F., M. Bravo, J. De Brito, 2012. Concrete with used tire rubber aggregates: Mechanical performance. *ACI Mater. J.* 109(3), 283-292.

[16] Bravo, M., J. de Brito, 2012. Concrete made with used tyre aggregate: durability-related performance. *J. Clean. Prod.* 25, 42-50.

[17] Alves, A. V., T. F. Vieira, J. de Brito, J. R. Correia, 2014. Mechanical properties of structural concrete with fine recycled ceramic aggregates. *Constr. Build. Mater.* 2014. 64, 103-113.

[18] Vieira, T., A. Alves, J. de Brito, J. R. Correia, R. V. Silva, 2015. Durability-related performance of concrete containing fine recycled aggregates from crushed bricks and sanitary ware. *Mater. Design.* 90, 767-776.

[19] Belin, P., G. Habert, M. Thiery, N. Roussel, 2014. Cement paste content and water absorption of recycled concrete coarse aggregates. *Mater. Struct.* 47(9), 1451-1465.

[20] Pepe, M., 2015. A conceptual model for designing recycled aggregate concrete for structural applications, Department of Civil Engineering - University of Salerno, Italy, 178 pages.

[21] Silva, R. V., J. de Brito, R. K. Dhir, 2015. Establishing a relationship between the modulus of elasticity and compressive strength of recycled aggregate concrete. *J. Clean. Prod.* 112, 2171-2186.

[22] Silva, R. V., J. de Brito, R. K. Dhir, 2015. Prediction of the shrinkage behaviour of recycled aggregate concrete: A review. *Constr. Build. Mater.* 77, 327-339.

[23] Olorunsogo, F. T., N. Padayachee, 2002. Performance of recycled aggregate concrete monitored by durability indexes. *Cement and Concrete Res.* 32(2), 179-185.

[24] Shima, H., H. Tateyashiki, R. Matsuhashi, Y. Yoshida, 2005. An advanced concrete recycling technology and its applicability assessment by the input-output analysis. *J. Adv. Concr. Technol.* 3(1), 53-67.

[25] Akbarnezhad, A., K. C. G. Ong, M. H. Zhang, C. T. Tam, T. W. J. Foo, 2011. Microwave-assisted beneficiation of recycled concrete aggregates. *Constr. Build. Mater.* 25(8), 3469-3479.

[26] Tam, V. W. Y., C. M. Tam, K. N. Le, 2007. Removal of cement mortar remains from recycled aggregate using pre-*soaking approaches. Resour. Conserv. Recy.* 50(1), 82-101.

[27] Tsujino, M., T. Noguchi, M. Tamura, M. Kanematsu, I. Maruyama, 2007. Application of conventionally recycled coarse aggregate to concrete structure by surface modification treatment. *J. Adv. Concr. Technol.* 5(1), 13-25.

[28] Soares, D., J. de Brito, J. Ferreira, J. Pacheco, 2014 Use of coarse recycled aggregates from precast concrete rejects: Mechanical and durability performance. *Constr. Build. Mater.* 71, 263-272.

[29] Pedro, D., J. de Brito, L. Evangelista, 2014. Performance of concrete made with aggregates recycled from precasting industry waste: influence of the crushing process. *Mater. Struct.* 48, 1-14.

[30] Fathifazl, G., A. Abbas, A. G. Razaqpur, O. B. Isgor, B. Fournier, S. Foo, 2009. New mixture proportioning method for concrete made with coarse recycled concrete aggregate. *J. Mater. Civil Eng.* 21(10), 601-611.

[31] Sato, R., I. Maruyama, T. Sogabe, M. Sog, 2007. Flexural behavior of reinforced recycled concrete beams. *J. Adv. Concr. Technol.* 5(1), 43-61.

[32] Evangelista, L., J. de Brito, 2014. Concrete with fine recycled aggregates: A review. *Eur. J. Environ. Civ. En.* 18(2), 129-172.

[33] Silva, A., R. Neves, J. de Brito, 2014. Statistical modelling of carbonation in reinforced concrete. *Cement Concrete Comp.* 50, 73-81.

[34] Silva, R. V., J. de Brito, R. Neves, R. Dhir, 2015. Prediction of chloride ion penetration of recycled aggregate concrete. *Materials Research.* 18(2), 427-440.

[35] Silva, R. V., J. de Brito, R. K. Dhir, 2014. The influence of the use of recycled aggregates on the compressive strength of concrete: a review. *Eur. J. Environ. Civ. En.* 19(7), 825-849.

[36] Silva, R. V., J. de Brito, R. K. Dhir, 2015. Comparative analysis of existing prediction models on the creep behaviour of recycled aggregate concrete. *Eng. Struct.* 100, 31-42.

[37] Silva, R. V., J. de Brito, R. K. Dhir, 2015. Tensile strength behaviour of recycled aggregate concrete. *Constr. Build. Mater.* 83, 108-118.

[38] Silva, R. V., R. Neves, J. de Brito, R. K. Dhir, 2015. Carbonation behaviour of recycled aggregate concrete. *Cement Concrete Comp.* 62, 22-32.

[39] Silva, R.V., A. Silva, J. de Brito, R. Neves, 2015. Statistical modelling of carbonation in concrete incorporating recycled aggregates. *J. Mater. Civil Eng.* 28(1).

[40] Soares, D., J. de Brito, J. Ferreira, J. Pacheco, 2014. In situ materials characterization of full-scale recycled aggregates concrete structures. *Constr. Build. Mater.* 71, 237-245.

[41] Li, W., J. Xiao, Z. Sun, S. Kawashima, S. P. Shah, 2012. Interfacial transition zones in recycled aggregate concrete with different mixing approaches. *Constr. Build. Mater.* 35, 1045-1055.

[42] Liu, Q., J. Xiao, Z. Sun, 2011. Experimental study on the failure mechanism of recycled concrete. *Cement Concrete Res.* 41(10), 1050-1057.

[43] Tam, V. W. Y., D. Kotrayothar, and J. Xiao, 2015. Long-term deformation behaviour of recycled aggregate concrete. *Constr. Build. Mater.* 100, 262-272.

[44] Xiao, J., T. Ding, 2013. Research on recycled concrete and its utilization in building structures in China. *Frontiers of Structural and Civil Engineering.* 7(3), 215-226.

[45] Xiao, J., B. Lei, C. Zhang, 2012. On carbonation behavior of recycled aggregate concrete. *Science China-Technological Sciences.* 55(9), 2609-2616.

[46] Xiao, J., W. Li, D. J. Corr, S .P. Shah, 2013. Effects of interfacial transition zones on the stress–strain behavior of modeled recycled aggregate concrete. *Cement Concrete Res.* 52, 82-99.

[47] Xiao, J. Z., J. B. Li, C. Zhang, 2006. On relationships between the mechanical properties of recycled aggregate concrete: An overview. *Mater. Struct.* 39(6), 655-664.

[48] Poon, C. S. Kou, 2010. Effects of fly ash on mechanical properties of 10-year-old concrete prepared with recycled concrete aggregates. *2nd International Conference on Waste Engineering Management, ICWEM 2010*. RILEM Publications SARL: Shanghai, China, 46-59.

[49] Kou, S. C., C. S. Poon, 2012. Enhancing the durability properties of concrete prepared with coarse recycled aggregate. *Constr. Build. Mater.* 35, 69-76.

[50] Kou, S. C. C. S. Poon, 2008. Mechanical properties of 5-year-old concrete prepared with recycled aggregates obtained from three different sources. *Mag. Concrete Res.* 60(1), 57-64.

[51] Poon, C. S., Z. H. Shui, L. Lam, 2004. Effect of microstructure of ITZ on compressive strength of concrete prepared with recycled aggregates. *Constr. Build. Mater.* 18(6), 461-468.

[52] Xiao, J. Z., W. G. Li, C. S. Poon, 2012. Recent studies on mechanical properties of recycled aggregate concrete in China-A review. *Science China - Technological Sciences*. 55(6), 1463-1480.

[53] Li, X., 2008. Recycling and reuse of waste concrete in China Part I. Material behaviour of recycled aggregate concrete. *Resour. Conserv. Recy.* 53(1-2), 36-44.

[54] Pedro, D., J. de Brito, L. Evangelista, 2014. Influence of the use of recycled concrete aggregates from different sources on structural concrete. *Constr. Build. Mater.* 71, 141-151.

[55] Li, X., 2009 Recycling and reuse of waste concrete in China Part II. Structural behaviour of recycled aggregate concrete and engineering applications. *Resour. Conserv. Recy.* 53(3), 107-112.

[56] Xiao, J., M. Tawana, X. Huang, 2012 Review of studies on structural performance of recycled aggregate concrete in China. *Science China - Technological Sciences*. 55(10), 2727-2739.

[57] Malešev, M., V. Radonjanin, S. Marinković, 2010. Recycled concrete as aggregate for structural concrete production. *Sustainability*. 2(5), 1204-1225.

[58] González-Fonteboa, B., F. Martínez-Abella, 2007. Shear strength of recycled concrete beams. *Constr. Build. Mater.* 21(4), 887-893.

[59] Ajdukiewicz, A. B., A. T. Kliszczewicz, 2007. Comparative tests of beams and columns made of recycled aggregate concrete and natural aggregate concrete. *J. Adv. Concr. Technol.* 5(2), 259-273.

[60] Corinaldesi, V., V. Letelier, G. Moriconi, 2011. Behaviour of beam-column joints made of recycled-aggregate concrete under cyclic loading. *Constr. Build. Mater.* 25(4), 1877-1882.

[61] Xiao, J., Y. Sun, H. Falkner, 2006. Seismic performance of frame structures with recycled aggregate concrete. *Eng. Struct.* 28(1), 1-8.

[62] Wang, C., J. Xiao, 2013. Study of the seismic response of a recycled aggregate concrete frame structure. *Earthq. Eng. Eng. Vib.* 12(4), 669-680.

[63] Pacheco, J., J. de Brito, J. Ferreira, D. Soares, 2015. Destructive Horizontal Load Tests of Full-Scale Recycled Aggregate Concrete Structures. *ACI Struct. J.* 112(6), 815-826.

[64] Pacheco, J., J. de Brito, J. Ferreira, D. Soares, 2016. Dynamic characterization of full-scale structures made with recycled coarse aggregates. *J. Clean. Prod.* in press.

[65] Pacheco, J., J. de Brito, J. Ferreira, D. Soares, 2015. Flexural load tests of full-scale recycled aggregates concrete structures. *Constr. Build. Mater.* 101, 65-71.

[66] Fonseca, N., J. de Brito, and L. Evangelista, 2011, The influence of curing conditions on the mechanical performance of concrete made with recycled concrete waste. Cement and Concrete Composites, 2011. 33(6): pp. 637-643.

[67] Amorim, P., J. de Brito, and L. Evangelista, Concrete made with coarse concrete aggregate: Influence of curing on durability. *ACI Mat. J.*, 2012. 109(2): pp. 195-204.

[68] EN-1990, 2002. Eurocode 0: Basis of structural design. *Comité Européen de Normalisation (CEN):* Brussels, Belgium.

[69] EN-1992-1-1, 2008. Eurocode 2 - Design of concrete structures: Part 1-1: General rules and rules for buildings. *Comité Européen de Normalisation (CEN):* Brussels, Belgium, 259 pages.

[70] ACI318-14, 2014. Building Code Requirements for Structural Concrete. *The American Concrete Institute:* Farmington Hills, Michigan, United States of America

[71] ASCE/SEI 7-10, 2013. Minimum design loads for buildings and other structures. American Society of Civil Engineers: Virginia, United States of America, 650 pages.

[72] Nowak, A.S., K.R. Collins, 2000. Reliability of Structures. McGraw-Hill: Singapore.

[73] ISO2394, 1998. General principles on reliability for structures. *Comité Européen de Normalisation (CEN):* Brussels, Belgium, 76 pages.

[74] MacGregor, J. G., 1976. Safety and limit states design for reinforced concrete. *Can. J. Civ. Eng.* 3, 484-513.

[75] Ferry-Borges, J., M. Castanheta, 1983. Structural safety - Third edition. Portuguese Laboratory of Civil Engineering: Lisbon, Portugal, 327 pages.

[76] Cornell, C. A., 1969. A probability-based structural code. *Journal of the American Concrete Institute.* 66(12), 974-985.

[77] Ellingwood, B., T. V. Galambos, J. G. MacGregor, C. A. Cornell, 1980. Development of a probability based load criterion for American National Standard A58. *NBS Special Publication 577, N.B.o. Standards*: Washington DC, United States of America.

[78] Rackwitz, R., B. Fiessler, 1978. Structural reliability under combined random load sequences. *Comput. Struct.* 9, 489-494.

[79] Ditlevsen, O., 1979. Generalized Second Moment Reliability Index. *J. Struct. Mech.* 7(4): pp. 435-451.

[80] Ditlevsen, O., 1979. Narrow Reliability Bounds for Structural Systems. *J. Struct. Mech.* 7(4): pp. 453-472.

[81] Madsen, H. O. K., S.; Lind, N. C., Methods of Structural Safety. 1986, New Jersey, United States of America: Dover Publications.

[82] Haldar, A. and S. Mahadevan, Probability, reliability, and statistical methods in engineering design. 1999: Wiley & Sons, 305 pages.

[83] Lemaire, M., Structural reliability, ed. J. W. Sons. 2009, Hoboken, New Jersey, United States of America, 504 pages.

[84] Szerszen, M., A. S. Nowak, 2003. Calibration of Design Code for Buildings (ACI 318): Part 2—Reliability Analysis and Resistance Factors. *ACI Struct. J.* 100(3), 383-391.

[85] Ditlevsen, O., H. O. Madsen, 2005. Structural Reliability Methods. *Department of Mechanical Engineering, Technical University of Denmark*, 327 pages.

[86] Rackwitz, R., 1984 Discussion to "Fast probability integration by three-parameter normal tail approximation." *Struct. Saf.* 2(2), 159-160.

[87] Thoft-Christensen, P., M. J. Baker, 1982. Structural reliability theory and its applications. *Springer.*

[88] Benjamin, J. R. C., C. Allin, 1970. Probability, Statistics and Decision for Civil Engineers: *McGraw Hill*, 684 pages.

[89] Mália, M., J. de Brito, M. D. Pinheiro, M. Bravo, 2013. Construction and demolition waste indicators. *Waste Manage. Res.* 31(3), 241-55.

[90] Agrela, F., M. Sánchez de Juan, J. Ayuso, V. L. Geraldes, J. R. Jiménez, 2011. Limiting properties in the characterisation of mixed recycled

aggregates for use in the manufacture of concrete. *Constr. Build. Mater.* 25(10), 3950-3955.

[91] Abbas, A., G. Fathifazl, B. Fournier, O. B. Isgor, R. Zavadil, A. G. Razaqpur, S. Foo, 2009. Quantification of the residual mortar content in recycled concrete aggregates by image analysis. *Mater. Charact.* 60(7), 716-728.

[92] Juan, M. S., P. A. Gutiérrez, 2009. Study on the influence of attached mortar content on the properties of recycled concrete aggregate. *Constr. Build. Mater.* 23(2), 872-877.

[93] Butler, L., Evaluation of Recycled Concrete Aggregate Performance in Structural Concrete, 2012. Department of Civil Engineering, University of Waterloo: Ontario, Canada, 475 pages.

[94] Henry, M., K. Hagiwara, T. Nishimura, Y. Kato, 2011. Effect of recycled aggregate quality on variation and estimation of concrete strength. *Proceedings of the Japan Concrete Institute*. 33(1), 1535-1540.

[95] Butler, L., J. S. West, S. L. Tighe, 2012. Effect of Recycled Concrete Aggregate Properties on Mixture Proportions of Structural Concrete. *Journal of the Transportation Research Board*. 2290, 105-114.

[96] Pepe, M., R. D. Toledo Filho, E. A. B. Koenders, E. Martinelli, 2016. A novel mix design methodology for Recycled Aggregate Concrete. *Constr. Build. Mater.* 122, 362-372.

[97] Xiao, J., J. Li, C. Zhang, 2005. On statistical characteristics of the compressive strength of recycled aggregate concrete. *Struct. Conc*. 6(4), 149-153.

[98] Etxeberria, M., E. Vazquez, A. Mari, M. Barra, 2007. Influence of amount of recycled coarse aggregates and production process on properties of recycled aggregate concrete. *Cement Concrete Res.* 37(5), 735-742.

[99] Xiao, J., K. Zhang, Q. Xie, 2016. Reliability analysis for flexural capacity of recycled aggregate concrete beams. Structural Engineering International. 26(2), 121-129.

[100] Breccolotti, M., A. L. Materazzi, 2010. Structural reliability of eccentrically-loaded sections in RC columns made of recycled aggregate concrete. *Eng. Struct.* 32(11), 3704-3712.

[101] Rackwitz, R., Statistical control in concrete structures. Specialized course - C4. 1973, Portuguese Laboratory of Civil Engineering: Lisbon, Portugal.

[102] Nowak, A. S. and M. Szerszen, 2013. Calibration of Design Code for Buildings (ACI 318): Part 1—Statistical Models for Resistance. *ACI Struct. J.* 100(3), 377-382.

[103] FIB, Bulletin n. 65, 2010. Model Code 2010 final draft. Volume 1. *FIB*: Lausanne, Switzerland.

[104] FIB, Bulletin n. 66, 2010. Model Code 2010 final draft. Volume 2. *FIB*: Lausanne, Switzerland.

[105] Rusch, H., 1969 Statistische analyse der Betonfestigkeit. *Deutscher Ausschuss fur Stahlbeton*: Berlin, Germany [Statistical analysis of concrete strength].

[106] ACI 214R-11, 2011. Guide to evaluation of strength test results of concrete. *The American Concrete Institute:* Farmington Hills, Michigan, 20 pages.

[107] Bartlett, F. M., G. M. James, 1996. Statistical Analysis of the Compressive Strength of Concrete in Structures. *ACI Mat. J.* 93(2), 158-168.

[108] Xiao, J., W. Li, Y. Fan, X. Huang, 2012. An overview of study on recycled aggregate concrete in China (1996–2011). *Constr. Build. Mater.* 31(0), 364-383.

[109] FIB, Bulletin n. 70: Code-type models for concrete behaviour, 2013. Lausanne, Switzerland, 196 pages.

[110] Kiureghian, A. D., O. Ditlevsen, 2009. Aleatory or epistemic? Does it matter? *Struct. Saf.* 31(2), 105-112.

[111] Breccolotti, M. A. L. Materazzi, 2013. Structural reliability of bonding between steel rebars and recycled aggregate concrete. *Constr. Build. Mater.* 47(0), 927-934.

[112] Woerner, J. D. and P. Moerland, 1998. Safety of RAC-Beams Designed According to Eurocode 2 Provisions. *Proceedings of the International Symposium on Sustainable construction: Use of Recycled Concrete Aggregate*. London, United Kingdom.

[113] González-Taboada, I., B. González-Fonteboa, F. Martínez-Abella, J. L. Pérez-Ordóñez, 2016. Prediction of the mechanical properties of structural recycled concrete using multivariable regression and genetic programming. *Constr. Build. Mater.* 106, 480-499.

[114] Ellingwood, B., T. V. Galambos, 1982. Probability-based criteria for structural design. *Struct. Saf.* 1(1), 15-26.

[115] JCSS, 2001. Probabilistic Model Code, in Part 1 - Basis of Design. *The Joint Committee of Structural Safety.*

[116] JCSS, 2001. Probabilistic Model Code, in Part 2: Load Models. *The Joint Committee of Structural Safety.*

[117] JCSS, 2001. Probabilistic Model Code, in Part 3: Material properties. *The Joint Committee of Structural Safety.*

[118] Mirza, S. A., J. G. MacGregor, M. Hatzinikolas, 1979. Statistical Descriptions of Strength of Concrete. *ASCE J. Struct. Div.* 105(6), 1021-1037.

[119] Mirza, S. A., J. G. MacGregor, 1979. Variability of Mechanical Properties of Reinforcing Bars. *ASCE J. Struct. Div.* 105(5), 921-937.

[120] Mirza, S. A., J. G. MacGregor, 1979. Variations in Dimensions of Reinforced Concrete Members. *ASCE J. Struct. Div.* 105(4), 751-766.

Sentler, L., 1976. Live load surveys: a review with discussions, Lund Institute of Technology: Lund, Sweden, 135 pages.

In: Construction Projects
Editor: Kimberly Hall
ISBN: 978-1-53610-742-5

Chapter 2

KNOWLEDGE MANAGEMENT MODEL FOR PROJECT MANAGEMENT: KM+PM$_{TIC}$

***Flor Nancy Díaz de Piraquive*[1], *Oscar Sanjuán Martínez*[2], *Elena Verdú Pérez*[2,*] *and Rubén González Crespo*[2]**

[1]Universidad Católica de Colombia, Colombia
[2]Universidad Internacional de la Rioja - UNIR, Spain

ABSTRACT

This research analyses how business processes of a company are related to the processes of projects, as the latter generally concentrate on the optimal joint to initiate, plan, execute, control and close a project, leaving aside the collaborative learning environment, which can benefit the development of the projects, and wasting technological tools provided by the use and appropriation of ICTs (Information and Communication Technologies) for knowledge sharing and transfer. Therefore, this research proposes to deploy a model of Knowledge Management for Project Management KM+PM$_{TIC}$ in order to facilitate best practices in the implementation of future projects, through the use of techniques and tools that facilitate knowledge creation, sharing and transfer, and supported by information technologies and communications. The final aim is to

* Corresponding Author address Email: elena.verdu@unir.net.

optimize the processes used in the development of projects from beginning to end, satisfying time restriction, costs, scope, quality, risk and use of resources, and motivating the project team members to work in collaborative environments that optimize the organization, letting them benefit from the experiences, lessons learned, best practices, generate new knowledge and increase the percentage of successfully completing projects.

Keywords: knowledge management, project management, collaborative learning, management facilitators, management model

INTRODUCTION

Project Management can be defined as the art that seeks, through innovative techniques and practices, to successfully implement projects from inception to completion of their activities.

Do managers, from a certain country or region, know how many additional costs (for unjustified delays, negligence or poor quality work) will there be for not knowing or being aware of the techniques of Project Management? The answer to this question is not encouraging. In fact, it would cause astonishment amongst people. People from developing countries or regions would be even more astonished, as investment resources are scarce. Unfortunately public and private companies have not understood the problem that occurs in Project Management for deficiency of an organizational culture where the team of the project knows and understands the importance of working with techniques and tools to facilitate creation, sharing and transfer of knowledge, to support the optimization of scarce resources allocated for projects.

Linked to the concept of Project Management, Knowledge Management concept has to be included, which for any organization represents the process that enables recovering, cataloging, storing and distribution of the contribution made by each professional for an activity during project development. Within the project environment, Knowledge Management is the dimension of the project where activities to improve future projects are done, in order to help implement solutions that already paid off and are already proven.

Basically, any good idea into a project has one of the following business objectives: improve efficiency ("doing things optimally"); improve effectiveness ("Doing the right things"); innovating a product or service

("Doing something new"). Therefore, the Project Manager is responsible for the Knowledge Management process throughout the project, more specifically in the closing phase of the project. The project manager and his team must ask: What went well? What was the best choice among the alternatives we had in that situation and circumstance of the project? What innovations can bring about the construction of a deliverable? Well, to answer the above questions, we should revise certain characteristics of human behavior that enable or encourage the sharing and transfer of knowledge in an organization to learn from mistakes made in the past. These are:

- First, knowledge is shared on a voluntary basis; you cannot force people to share or transfer it. What should the project manager do about this? Create a collaborative environment where team members feel comfortable sharing, transferring and teaching what they know, and last but not least, which helps them understand how good this is for the organization.
- Second, people know more than what they can spread or document. What should the project manager do regarding this? He should recognize the production process, classification and storage of information as part of the work involved in the project. In case the team members are not aware of how to do it, he should train and teach them how to do so.

The project closure is the phase in which the project manager has the highest probability of contributing to Knowledge Management. This phase is not a "moment" of the project, but a process. In this process, the project manager gathers all the information produced, catalogs and stores it in a way that it is easy to search and retrieve in the future. It would be good that the manager paid enough attention to Knowledge Management in the early phases of the project, but the closing phase of the project is the best opportunity to do it, because all knowledge has already been produced and the team involved in the project is still present.

The Knowledge Management process is more specifically called (sometimes) Organizational Memory. You can consider the project manager as the principal responsible for enriching the memory of the organization in the process of developing the project, documenting and classifying what they learned for the benefit of future projects. The project manager must take advantage of new Information and Communication Technologies (ICT) and promote the use and appropriation of the facilitators of knowledge creation and sharing, to optimize resources, improve quality, efficiency and productivity. The project manager must make clear to the team, the advantages gained by implementing Knowledge Management strategies and by the use of techniques and tools of Project Management, which are combined to reach an efficient integration of all processes for execution thereof.

The research described in this chapter has resulted in a model of Knowledge Management for Project Management, in order to facilitate best practices in the implementation of future projects, from the use and appropriation of Knowledge Management facilitators based on new technologies of information and communication.

THEORETICAL FOUNDATIONS

Today Knowledge Management is used by many organizations as a competitive strategy for positioning, stay afloat and be recognized in the market as a company that achieves its objectives and delivers quality products and services, as well as innovative. In recent years, Knowledge Management has become a valuable and important resource, therefore it must be lead, managed, controlled, shared and transferred (Díaz, Medina & Joyanes, 2013).

Knowledge has always been present throughout history, from ancient Greece, through the Middle Ages, Renaissance, Rationalism and Contemporary Age up to modern authors. According to Domingo Valhondo in his book titled “Knowledge Management Myth to Reality” (Valhondo, 2003), in the 90s decade with the arrival of Internet, Knowledge Management began to evolve and become popular at organizational level, capturing the attention of large companies located in the United States, Europe and Japan. Knowledge Management was defined as a business term after Tom Stewart published “BrainPower” in Fortune magazine (Stewart, 1991).

Knowledge Management Models for Project Management

Among the most prominent models known, and which have been analyzed for the construction of the model of this research, we found:

- *Spiral Model*: It proposes a model for the process of creating knowledge, describing it as a four-phase process, in a spiral shape

(Nonaka & Takeuchi, 1999; Nonaka, Reinmoeller & Senoo, 1999). In this model, the authors speak of tacit knowledge and explicit knowledge, separating each one in two pillars.

- *Model integration with the KM PM for project success*: In this model, the authors (Yeong & Lim, 2010) make an integrated proposal taking into account factors such as culture, processes and technology, which clearly impact Knowledge Management (KM) and Project Management (PM).
- *KM role model of organizations in the PM of complex systems*: According to Gudi & Becerra-Fernandez (2006), this model considers three main aspects. These are: the conclusions of research reports that determine the errors in projects; the conditions that cause the errors that existed before the failure; and the ability to better understand the nature of risk, to reduce or remove the risk of failure. Risk reduction and prevention of failures during system development focuses on dynamic aspects of the PM.
- *Performance Evaluation Model of KM in PM*: The authors of this theory (Humaidi & Asarani, 2012) indicate that the model contains the following elements: leadership style; personality of the project manager; teamwork based on the implementation capacity of the project team members and suitability for the tasks they are assigned, to efficiently and effectively articulate the project life cycle, where the project tasks are structured in accordance with the processes; knowledge, selected as a moderating variable in this framework, which has its importance highlighted, to be shared by project stakeholders; key performance indicators (KPI), measurable with reference to the actual performance and what is expected in the result.
- *Knowledge Management Model for Project Management*: The model interrelates the sharing of individual knowledge with the organization, starting from the implementation of the processes of knowledge combined with project processes, which results in knowledge transfer and the application of best practices while implementing projects (Díaz, Medina & González, 2015).

In summary, the models analyzed conclude and coincide with the application of knowledge transfer in the development of processes that leverage best practices in inception, planning, execution, monitoring and control, and closing projects.

Knowledge Management facilitators

Below there is a list of techniques and tools that ensure the creation, sharing and transfer of knowledge, which purpose is to improve communication and professional relationships, uncover and dissolve hidden conflicts, increase the effectiveness in achieving results and promote emotional generative states that facilitate the sharing and transfer of knowledge, from a collaborative environment that uses tools and best practices.

In accordance with the theory, it can be said that a facilitator has characteristics that identify him as a leader, from which fulfills the function of providing elements that contribute to the achievement or enforcement of a goal. For this reason we will describe below some techniques or mechanisms that facilitate the creation or generation (and therefore the comparison and transfer) of knowledge in the framework of a project development, resulting in the generation of knowledge in the Organization.

- *E-Learning Systems*: They strengthen the interactivity, let the person who uses the tool be the protagonist of his own learning. They can be defined as all those methodologies, strategies or learning systems that use digital technology and/or communication mediated by computers to produce, transmit, distribute and organize knowledge among individuals, communities and organizations (Bernandez, 2007). E-learning is mainly an electronic environment for distance or virtual learning, where you can interact via Internet.
- *Weblogs (Blogs):* Also called log, it is a web page with notes dated in reverse chronological order so that the most recent entry is the one that first displays. A weblog is foremost, a free form of expression, creation and sharing of knowledge (Santamaría, 2007).
- *Technological Surveillance:* it is based on a group of people dedicated to see and analyze the various sources of information of the Organization. It strengthens the project management in the organizations, since it captures and analyzes important scientific and technological information as well as connected information such as economical behavior of the sector, competition, market and environment, providing aspects that allow a strategic decision-making during the planning, execution and monitoring of projects.

- *Wikis:* Collaborative website brought forward by the collective work of many authors. It allows anyone to add, edit and delete content, although this content has been created and shared by other authors. It is linked to concepts of collaboration and community (Díaz, Medina & Joyanes, 2013).
- *Expert judgment*: According to Esterkin (2008), the judgment of experts is "a set of opinions and judgments that professional experts can provide regarding an industry or discipline, related to the project that is being executed." This information can be obtained within or outside the organization.
- *Lessons learned*: According to Medellin (2006), they are defined as the set of knowledge resulting from the implementation of a project, a result of events, experiences, actions or omissions allocated during its development and which allow people to acquire a set of teachings, based on a process of continuous learning experiences or reports of the project management.
- *Communities of Practice*: These are defined as groups of people who share their expertise and their passion about some issues and interact to continue learning about the matter.
- *Coaching*: It is defined as the assignment of the role of "coach" to the executive or manager of a certain group. He has to take care of the subordinates in a day-to-day basis relationship between bosses and subordinates.
- *Mentoring*: It is considered a personal learning process, where persons are responsible for their own personal and professional development. Thus, a personalized relationship between mentor and mentee is established.
- *Storytelling*: According to ADECEC, an association of consulting firms in public relations and communication, storytelling is defined as oral or written constructions involving two or more people who interpret a past experience (ADECEC, 2013).

Project Management has been faced with many challenges generated in response to technological changes of recent decades. These changes have also been supporting projects, research, where the teaching has incorporated the use of ICT to different learning models. The development of ICT has strengthened the way of imparting and acquiring knowledge, as it has been taken to other training scenarios: from written or printed work, documents, exhibits on paper, among others, to computers, digital documents, email, chats, PDAs and other more agile and effective means, giving the opportunity to comment, contribute and share knowledge in collaborative environments (Myriam, 2007).

To ensure that the collaborative virtual tools can support the project management, there must be commitment from the top Management, as the creation of collaborative environments must be encouraged and promoted from the inside of the Organization, as they will facilitate the interaction between participants and the projects.

Formulation of the Problem and Objectives

In the 21st century companies that are managed with progressive criteria (and not principles aimed at mere survival), have found in Project Management and in the office that administers it (PMO: Project Management Office), support to business growth based on the principle that "companies that do not grow, dig their own grave." Projects are key to take businesses to assume a real competitive role and should be the foundation for defining a model of Knowledge Management to optimize Project Management for the company.

The Knowledge Management on the basis of the development of projects is then the result of the detailed assessment of the variables scope, time, cost, quality, risks, and use of resources. It is also the result of the development, use and appropriation of the new information and communication technologies, which should be part of the facilitators for the generation and transfer of new knowledge among stakeholders in the project. These stakeholders are individuals, groups or organizations that can affect or be affected by the project.

To reach maximum leverage of a model of KM, it is necessary to determine the barriers that the methodologies that use these techniques, can present. Within the first findings there are management profiles that hinder growth and sustainability of enterprises. This has to do with the most common errors in the development of the investment and in the implementation of the planning and programming of projects. Studies of international expert entities such as Project Management Institute (PMI), McKinsey & Company, International Data Corporation (IDC), Gartner, John McManus and Chaos Knowledge Center (CKC), among others, share their analyses and conclusions applied in each one of the stages of the project life cycle, evidencing statistics with surprising results that lead us to wonder about the role and skills of the project manager, the work methodology used during the execution of it, risk management, team work, environmental factors and change management, among others.

The statistics of success of the projects of information technology (IT) still remain below the expectations of the market (Tan, 2011). Gartner, one of the largest groups of reflection on it, said "75% of all IT projects fail" (Hillman, 2010). The Standish Group International, Incorporated (2013) affirms that although statistics for the years 2004 to 2012 have improved, the problems are still about schedules, budget and the characteristics of the scope. The statistics for the year 2012 show a slight increase in the costs and increases in the time. The overruns rose from 56 per cent in 2004 to 59% in 2012. The increments of time also rose from 71% in 2010 to 74% in 2012, although the highest point in the excesses of time was 2004 (84%). The features and functions developed around the scope fell from 74 per cent of the specified requirements completed in 2010, to 69% in 2012. This could be a good sign because organizations spend more time focusing on the requirements that generate high value, instead of completing the 100% of the total requirements of the projects (Díaz, González & Medina, 2015).

The Project Management Institute collects the information provided by project managers and business analysts and says that 47% of unsuccessful projects fail to meet the needs for the lack of specification and clarity in the requirements (Smith, Bieg & Cabrey, 2014).

The similarities of these results are quite remarkable, taking into account that the contexts vary from country to country, industry and even the sector. The statistics evaluation of IT projects show that 30% of public projects oriented to the design of information systems tend to completely fail, 60% reported some kind of failure and only the minority ends up with success (Galunic & Rodan, 2007). Models, protocols, tools and services have been created around this industry, aimed at solving the problems of persons and entities that are involved in IT projects.

Another problem identified as a cause for failure of the projects is the governance. This plays a decisive role and is part of the fundamental criteria of failure. The governance of a project involves both: those involved in it and the policies, procedures, standards, responsibilities and authorities documented (Project Management Institute, 2013). If governance was supplemented with the skills of the project manager at the moment of evaluating the reasons for failure, the criteria for achieving success could be obtained in different aspects: acceptance of deliverables; management of incidents and problems arisen during the life cycle of the project (Project Management Institute, 2013) in relationship with the team and the stakeholders; communication; decision-making; the review process of the phases or plans in the review process of changes and risks; and, in general, the articulator process to align the interested stakeholders with the requirements of the project. It is important to recognize that the skills and competencies both, soft and hard, as the project manager, are attributes that facilitate the success of the projects. Soft skills are those attributes or characteristics of people that allow them to interact with others effectively, which generally focus on the work or certain aspects of this. Hard skills are related to the technical skills needed or acquired to perform certain tasks or functions that are achieved and developed through education and training.

In addition to counter the above, the report of CHAOS 2015, published by the Standish Group (Standish Group, 2015), applied to 50,000 projects around the world, states small improvements to those who implemented reengineering to their massive systems. It is there evidenced the current reality of the state of the IT industry. The statistics of project's termination status executed from 2011 to 2015 included in this report state that while 22% of projects failed in 2011, 19% did it in 2015. However, only 29% of projects are successful in 2015. Although it includes enhanced definition of successful projects, the results indicate that there is still work to be done towards the achievement of successful results in IT projects.

In the approach of critical variables supported by experts such as common reasons for failure in IT projects, these are considered to be:

- *Managerial stultification:* It refers to the managers concerned in maintaining their image, who use to reject any proposal that involves some degree of risk and therefore do not focus or interest in investment projects that stimulate growth.
- *Disregard of growth:* It is a common characteristic of businesses that are run with clear criteria of survival. This profile is in full wrong way with continuous research, design of new products and introduction of the services to new markets. These organizations simply resign themselves to what they have.
- *Lack of knowledge of the environment:* This occurs when the company does not have updated information of the market and the economic, social or political environment.

- *Lack of mission*: Top Management does not clearly address the specific mission of the company and other differentiating elements.
- *Limited research project*: Whenever the implementation of investment projects depends on the manager's intuition, the financial problem that occurs is usually focused on inefficient use of cash flow.
- *Inaccurate dimensioning of the scope, sizing, quality, time, and cost*: When sizing up the project, it is necessary to take into account the multiple factors involved in the development. Most determinants factors are time, resources, costs and processes definition.
- *Lack of communication between areas*: Communication is an essential mechanism in project management, which includes: the knowledge that is being communicated, who is communicating, the means of communication, and frequency of communication (Morreal, 2012).
- *Management and leadership skills:* In a world that is constantly changing, leaders with the ability to identify and adapt in the best possible way and in the shortest time to changes, are highly demanded.
- *Equipment integration:* Human resources integration and teamwork are factors that guarantee good synergy when developing the project, managing to accomplish individual and collective goals.
- *Change management*: One of the inevitable processes of daily life is the transition from one situation to another different one, creating what is known as change. An appropriate way to manage it is through the assessment and planning thereof, proposing alternatives that are easily understood by the project team members.
- *Inappropriate controls:* It occurs when there is no planning or control of the activities in the development phase of the project and the execution is continuously improvised.

According to Rodríguez (2005), other common causes of failures in IT projects are: deficient commitment of top management, users not getting involved, lack of technical knowledge within the team, deficiency of maturity or stability of the technology used in the project, bad relationships between phases or areas involved in the project, poor administrative and economic management, lack of supervision on the team and supervisors, scarce follow-up meetings and control.

According to Cerpa & Verner (2009), the causes that lead to failure in projects are: delivery date, the project was underestimated, the risk is neither re-evaluated nor controlled nor managed through the project, staff not feeling rewarded for working long hours. Other causes are: decisions delivered without the appropriate requirements information, inappropriate development methodology for the project, team motivation affected by aggressive schedules, changes in the scope during the execution of the project, negative impact of the implementation on the lives of the members of the team, insufficient staff to meet with the programming goals, late recruitment of staff to comply with an aggressive programming, and lack of time to collect the requirements by users/clients.

All the above mentioned reasons seem to suggest that PM is limited to be a system of implementation and control of projects, and it is not being considered as an enabler of a learning environment, where you can generate, share and transfer knowledge through the learning of best practices, lessons learned or other technological or organizational facilitators that optimize the implementation of future projects.

It is a challenge for the project manager not only to manage the efforts of his team around the financial and technological assets, but also to lead all team members and develop the capacity to influence through motivation, using the enabling tools to create a collaborative environment of learning that will lead to the success of the undertaking of projects and therefore, to the achievement of the objectives of the Organization.

The objective of this research is to formalize a model of Knowledge Management (KM) for Project Management (PM) for the use of techniques and facilitator tools for creation, sharing and transfer of knowledge, supported by the information and communication technologies that help processes used in the development of projects from its inception until the end, satisfying the time restriction, costs, scope, quality, risks and use of resources, and, motivating the members of the team to work in collaborative environments that optimize the benefits of the organization on the basis of the experiences, lessons learned and best practices.

METHODOLOGY

The methodology applied in this research first implies the documentary analysis related to physical or virtual writings and the review of cases documented in the context to which the research is addressed. After the information is selected and analyzed, it is organized and evaluated according to three phases or procedures of development:

In the first phase there is a theoretical analysis through a historical and descriptive review of knowledge and aspects that are relevant to the KM with emphasis in its conception and application in the development of projects. An analysis of bibliographic sources is made in printed and digital format. Help is obtained from web consultation observation, interview through forums, participation in virtual chat and academic forums and through the documentary analysis. Once this information is compiled, it is the moment where the theoretical framework is formed, with the vision of standardization of concepts towards KM.

In addition, there is another analysis that is done: a documentary analysis of the Project Management methodologies for the scope, objectives, structure and tools used in the processes of inception, planning, execution, monitoring, control and close. The purpose is identifying situations that could benefit from the application of KM techniques and principles, which lead to the creation, sharing and transfer of knowledge and continuous improvement.

In a second phase, an empirical method is used, based on the study of cases, identifying the organizational profile of the companies studied, the evolution and trajectory that led to the implementation of models of KM and the best practices obtained by it. According to Robert Yin (Yin, 1989) cited by Victor Medina (Medina, 2004), a case study is an empirical research that explores the contemporary phenomenon within its real context, when the borders between the phenomenon and the context are not evident and in which multiple sources of evidence are used.

In the third and last phase, a deductive method is used, on the basis of the theoretical background and analysis of cases of organizational models, with the purpose of identifying those facilitators of knowledge sharing, which will reflect on the processes of the project management. In this phase, it is also done the analysis of the KM technology that can be used from the era of Web 2.0, to formulate the organizational model of Knowledge Management for project management, focused on an approach to a collaborative environment that fosters the development of education and participation. This has often been called "collective learning." Then, the conclusions, reflections and suggestions that confront the contributions to the objective set out in the investigation are made (Figure 1).

RESULTS

After analyzing the guidelines and methodologies for project management, as well as the collaborative facilitators that optimize its processes, consequently here we find a description of the proposal for a model of Knowledge Management for Project Management KM+PM$_{TIC}$, supported by ICTs. The model aims at satisfying the restrictions and motivating the members of the team to work on collaborative environments, where you pick up the programming derived from the implementation in different scenarios and document the results, with the purpose of providing elements that facilitate the replica of the intervention in similar cases, optimizing the benefits of the organization on the basis of the experiences, lessons learned and best practices.

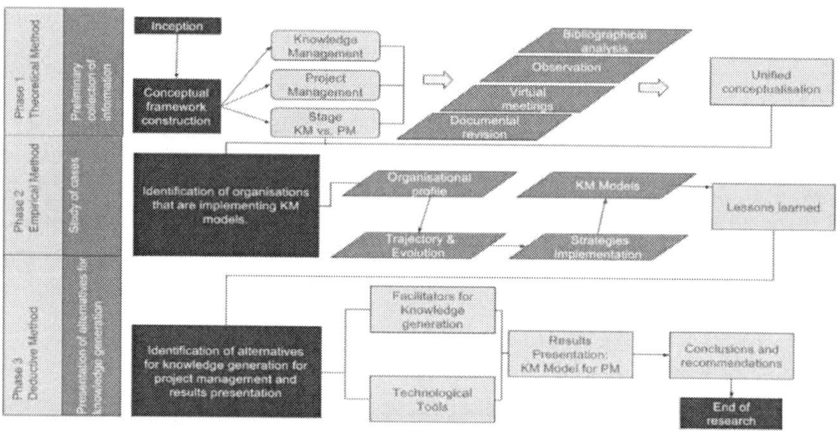

Figure 1. Phases of the investigation methodology.

In accordance with Richard Jaramillo Herrera (Jaramillo, 2012), before describing the model of Knowledge Management, it is important to identify its core principles to ensure the results. Firstly, this author proposes a collaborative organizational culture as the main basis for the model. A collaborative organizational culture is characterized by, among others, the following aspects: support to learning, commitment to the successes of others, dynamic interactions, democratic decision-making, etc. The change towards a corporate culture should be done involving the whole team in the process of change, asking them about the aspects of the existing culture that weaken the exchange of knowledge and enforcing collaboration, trust and inquiry. Other principles on which the model proposed by Jaramillo is based include motivations for knowledge sharing, confidence and trust between team members, top management backup, experiences recovery and continuous learning.

Presentation of the Model KM+PM_{TIC}

To start with the description of the model, we will begin by saying that within the management scope, the model refers to the process carried out by an administrative or management unit with the aim to control and organize the activities that have been entrusted. This way, the management model represents a set of procedures to be carried out to resolve an issue, to identify a project or manage a company or organization, i.e., a management model is a schema or framework of reference for the administration of an entity. The management models can be applied in both the corporate and private business and in the public administration.

The proposal for development of a model of Knowledge Management for the project management supported by KM+PM_{TIC} is based on:

- The Knowledge Management models analyzed, which have common elements related to the tacit and explicit knowledge and the way that knowledge is transferred, from the identification, evaluation, creation, externalization, appropriation and transfer.
- The guidelines of the Project Management Institute, described in the fifth edition of the PMBoK *(Project Management of Body of Knowledge)* (Project Management Institute, 2013), which establish five processes (initiation, planning, execution, monitoring and control, and closure) and ten different types of knowledge (integration, scope, schedule, cost, quality, human resources, risks, procurement and stakeholders).
- The facilitators of Knowledge Management, the Office of Knowledge Management and Intellectual Capital, analyzed in the theoretical foundation.

In order to reach the stated objective, the model will articulate the main facilitator of knowledge sharing, the Project Management Office and the structure of the intellectual capital of an organization, which supports the planning, implementation, monitoring, control and closure of projects, promoting learning scenarios, use and appropriation of information and communication technologies, resulting in collaborative work environments that leverage experience, lessons learned and best practices, breaking the myth that considers Project Management only as scenarios of control.

The model presents a general scheme in which two layers are described: the base or foundation and management (Figure 2). The goal is to develop activities that lead to the transfer of knowledge, from personal behavior and knowledge to collaborative, where the project strengthens and new knowledge is acquired as well as skills, which translate into higher profits for the organization.

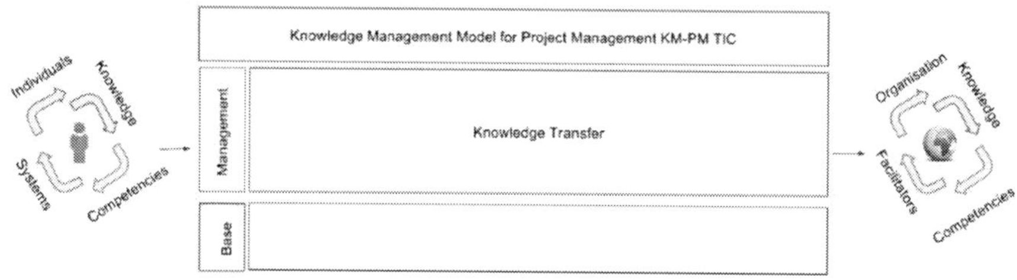

Figure 2. General Schema of Knowledge Management KM+PM$_{TIC}$ model.

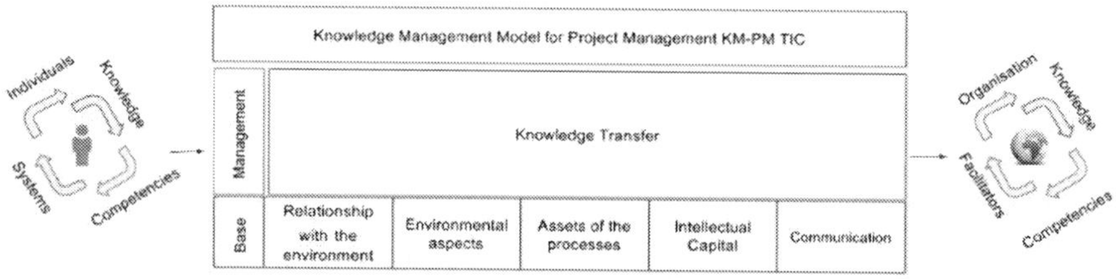

Figure 3. Description of the Base or Foundation of the Knowledge Management KM+PM$_{TIC}$ model.

In addition, the model is based on the structure of a Project Management Office (PMO), which through the characterization and categorization of the information generated by the project in their repository will support the entire management.

Layer Base or foundation of the Model KM+PMTIC

This layer is made up of five components (Figure 3): relationship with the environment, the environmental factors of the Organization, the assets of the Organization, intellectual capital and communication. These elements support the management layer and are kept in a virtuous cycle of knowledge of feedback and continuously updated processes both KM and PM, optimizing the development of the proposed model.

Relationship with the Environment

In order for the project to reach its objectives and obtain competitive advantages during the execution, it is needed to remain vigilant and be permanently identifying changes occurring in the environment. It has to be agile enough to alter strategies and plans, when difficulties arise. This component takes into consideration social and economic variables, fiscal and tax rules, political and economic framework, economic impact and the market sector where the project will be developed.

Environmental Factors of the Organization

These can have a major impact on the PM, even in environments that are relatively stable, with financial access, technology, and people. Projects succeed or fail according to the ability of the leaders to make the most of the available resources. Here it is important to consider the market conditions, risks, tolerance, human resources administration, and organizational structure, all for the purpose of strengthening the organizational culture, which is the core and most important factor to be considered in this component model.

According to David W. De Long and Liam Fahey (2000), the organizational culture plays an important role in sharing, ownership and knowledge transfer, as culture is responsible for conditional behaviors that facilitate or hinder the implementation of this practice. The author proposes four frames around the culture, these are:

- Culture conditions the creation and adoption of new knowledge;
- Culture determines the beliefs about the importance of knowledge;
- Culture intercedes in the relationship of levels of knowledge,
- Culture creates a context for social interaction.

It is of vital importance to the project to strengthen these elements.

No matter how efficient is a project manager or how carefully he planned a project, unforeseen environmental changes (disasters or economic changes) may affect the project at any stage of its implementation. Insofar as possible, the leaders of a process and project managers must be alert to the possibility of unforeseen environmental impacts and have contingency plans to ensure the least possible negative impact, for the success of the project (Project Management Institute, 2013).

Assets of the Processes within the Organization

They are composed of any or all of the assets relating to processes of the Organization in which the project is developed. These can be used to influence the success of the project. They can be grouped into two categories:

- Processes: consisting of plans, policies, procedures, certifications and guidelines both, formal or informal.
- Corporate knowledge base of the Organization, such as the lessons learned and historical information.

The assets of the organizational processes provide criteria and guidelines for adapting those processes to the specific needs of the project (Project Management Institute, 2013).

Intellectual Capital

Intellectual Capital is defined as a non-financial resource that allows the generation of responses to market needs and helps to exploit them. These resources are divided into three categories: Human Capital, Structural Capital and Relational Capital (Nonaka & Konno, 1998).

- Human capital: The capacities of individuals in an organization that are required to provide solutions to customers. For example, the individual and collective capacities, leadership, experience, knowledge, and special skills of the participants of the organization.
- Structural capital: The organizational capabilities required to respond to the requirements of the market. For example, patents, know-how, the secrets of business in the design of products and services, the accumulated knowledge and its availability, systems, methodologies and the culture of the organization.

- Relational capital: Depth (penetration), width (coverage), and profitability of organizational rights. For example, brands, consumers, the loyalty, reputation, channels and special contracts.

The intellectual capital enjoys a privileged place at the time of successful development of processes, to reach the objectives of the project, satisfying the restrictions related to the scope, time, cost, quality, risk and project management.

Communication

The purpose of the communication is to keep all stakeholders informed and aware of the development of the project and to exchange reports, recommendations, opinions and solutions.

According to Maria Luisa Sanchez Calero (Sanchez, 2005), there is no good external communication without good policies of internal communication; they are two sides of the same coin that are held together in his theoretical approach but that there is a dissociation in the practice of the company. This leads to consider that it is necessary to ensure communication within an organization, in such a way that this does not affect the external communication, since it is not a good practice to look at them separately.

Communication management is vital for companies and projects. According to the Project Management Institute (2013) communication management includes the processes required to ensure that the planning, collection, creation, storage, distribution, recovery management, control and monitoring of the project information is timely and appropriate. This leads to think that much of the success of a project focuses on communication management. Thanks to communication, teamwork is more efficient; generate a better working environment in which differences and arguments decrease significantly, thus achieving better results in different areas. This also creates synergies that make projects smoother and fluent, thus achieving the goals, as Bonilla Gutiérrez (1988) manifests, communication promotes the coordination of activities among the individuals involved, and enables the achievement of fixed goals. Accordingly, we can say that effective communication creates a link between the various project stakeholders, which have divergent cultural and organizational conditions, without becoming an element that threatens the implementation and result of a project.

Management Layer of the Model KM+PM$_{TIC}$

The management layer is composed of two major phases (Figure 4), the Knowledge Management and Project Management. These two phases are articulated with the facilitators of KM that promote the transfer of knowledge.

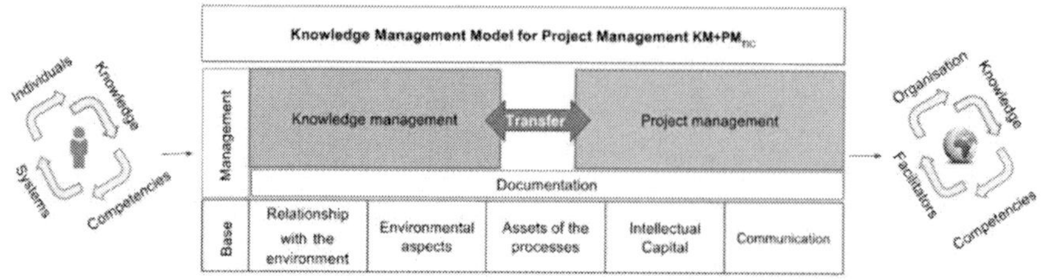

Figure 4. Description of the Management Layer of the KM+PM$_{TIC}$ model.

It is important to highlight that when speaking of KM in organizations, reference is being made to the process or the situation that enables knowledge or experiences transfer from a person or group of people to other persons belonging to the same company. Core components of KM are information and skills of the person or company and, if it has access to virtual collaborative tools, internal and external knowledge can be articulated to detect what happens in the competitive environment in which business is developed.

On the other hand, PM is a branch of administration science that studies the planning of the implementation of a project before its start, and the monitoring and control to measure the progress of the project, making decisions that are necessary to reach the objectives from the beginning and maintain performance levels expected. A correct and detailed definition of the project is essential for a correct planning, development, monitoring, control and close.

The two phases of the management layer must be supported in an area that guarantees an adequate structure of the documentation. In the proposed model, that area is defined in the Project Management Office PMO, as the "repository of information." All reports, studies, plans, contracts, business models, proceedings, incidents, acceptance testing, claims are classified, categorized, and guarded. In addition, changes and any evidence of the implementation, of all the processes in the project, from the point of view of trade, technical, administrative, financial, legal and learning are classified, categorized, and guarded too.

Below, processes of the phases of Knowledge Management and Project Management, proposed in the model, are described.

Phase of Knowledge Management

Within this phase, the model aims to develop the processes of identification, evaluation, creation, externalization, appropriation and transfer (Figure 5).

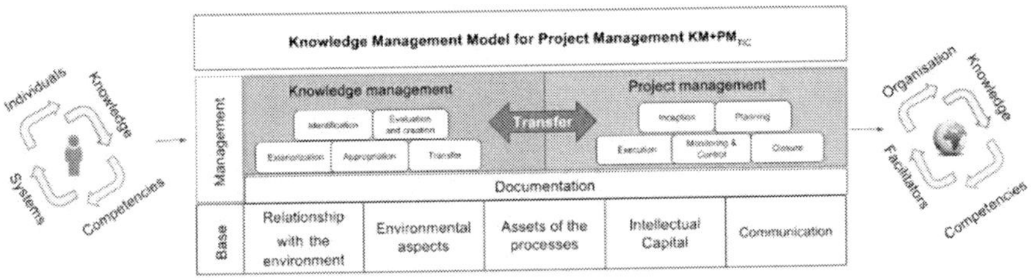

Figure 5. Processes of the phases of KM and PM of the $KM+PM_{TIC}$ model.

Identification

Identification is the process of locating cognitive models and assets, valuable to the project and for the organization. Assets refer to thoughts and actions that are owned by people who, according to their cognitive and innovation capacities, determine the new sources of knowledge. The sources of knowledge are generated through participation in internal activities such as projects or I&D, as well as through external resources such as training courses or sources in Internet (Jaramillo, 2012).

Evaluation

Evaluation is the process of selecting the model according to a criterion of interest. The criteria can be based not only on organizational criteria but also on individual or group criteria. According to (Jaramillo, 2012), it would be ideal that the people who participates in the identification of the model were trained and authorized to evaluate it, as this will make the task of selecting new models more effective.

Creation

The creation of new knowledge is conceived as the contribution the team does to the solutions, through ideals or visions that can serve as an incentive for innovation. The essence of innovation is rethinking of the world, according to an ideal or a particular vision. Creating new knowledge literally means to re-create the company, the project or the whole team, within a continuing process of the renewal of human talent and development of the Organization (Nonaka, Reinmoeller & Senoo, 1998; Díaz, Medina, González & Liberona, 2014).

According to Nonaka and Takeuchi (1999) there are three key features of the creation of knowledge:

- To express what is unspeakable, for which it relies mainly on the figurative language and symbolism. By means of the metaphors, people unify in a new way what you know and begin to express what they already know but cannot yet say.
- To disseminate the knowledge, since the personal knowledge of an individual must be shared with others. The knowledge can be crystallized in the group through the dialog, discussion, the exchange of experiences and the observation.
- The new knowledge is born between the haze of ambiguity and redundancy. Sometimes, the ambiguity is useful not only as a source of a new sense of direction, but also as a source of alternative meanings and a new type of thought (Nonaka & Konno, 1998; Nonaka, Reinmoeller & Senoo, 1998).

The creation of new knowledge is the product of a dynamic interaction between stakeholders, mainly between top management: front line employees and executives. The mid-level controls synthesize the tacit knowledge of both the top management and senior executives, make it explicit and incorporate it into new products and new technologies through the facilitators of KM (Díaz, Medina & González, 2014).

Externalization

Externalization is the core activity in the creation of knowledge. It consists in the conversion of tacit knowledge into explicit knowledge through the use of techniques and tools that facilitate its understanding. It takes the form of metaphors, analogies, hypotheses and models. Concepts are created through the deduction and induction, integrating it into the culture of the Organization (Nonaka & Konno, 1998). It is based on the articulation of tacit knowledge that passes to the Organization, involves techniques that help to express the intuitions and images of an individual as schemas, concepts and figurative language.

Appropriation

Appropriation is the strategy that seeks to promote the retention and ownership of knowledge, understanding by it, a process of intervention and comprehension of the relationships between the team involved in the project, use and appropriation of the techno-science and legal society represented by the client, built from active participation of the various groups of the project that generate knowledge. According to Lozano (2010), this process has the following characteristics: it is intentional; it is constituted by groups of experts in management and technology, generating mediations; it is a process where the manager and the team are empowered by knowledge; ownership is not alienation, it implies, even in the most asymmetrical relationships, the translation and assembling within the frames of reference of the participating groups. This understanding extends the dynamics of knowledge production beyond the synergies between the members of the project team; including the stakeholders. This enlargement integrates appropriation and innovation at the same level, under the principle of knowledge construction.

Lozano (2010) defends that the social appropriation of knowledge is the foundation of any form of innovation because knowledge is a complex construction that involves the interaction of different groups. Knowledge production is not a construction unrelated to society, as it develops within it, on the basis of their interests, codes and systems. On the other hand, innovation understood as the effective incorporation of social knowledge in the solution of problems, or in the establishment of new relations, is not more than the interaction between groups, artifacts, social cultures of experts and non-experts. The appropriation is not a passive reception; it always involves an interpretative exercise and the development of reflective practices. Therefore, the model for the construction of new knowledge and innovation will be supported with the facilitators of Knowledge Management, which are listed and are part of this research, according to the needs.

Tranference

Transference is the process that analyzes the opportunities for the transfer of knowledge by emphasizing the experiences, lessons learned and good practices. It also identifies items related to learning, improvement proposals, deployment and articulation of processes. These elements should promote the clarity and precision of the transfer of knowledge in itself, motivating the perception of salient aspects of the source of knowledge; the presence of the retention or motivation of who will receive that knowledge; and the positive features that can be seized in that context or environment where the project is developed. In this process, the knowledge is characterized for its storage, protection, transmission and use, both internally and externally. It is the technological window that will support the operation of the model.

Figure 6 shows the articulation of the processes of Knowledge Management, in relation to the stage of project management. The processes of the phase of KM flows into the time interacting with the processes of the phase of PM, giving as a result the transformation of the individual knowledge in communities of practice, collaborative learning and use of best practices.

Phase of Project Management

Within the phase of project management, the model aims to develop the processes of inception, planning, implementation, monitoring & control, and closure.

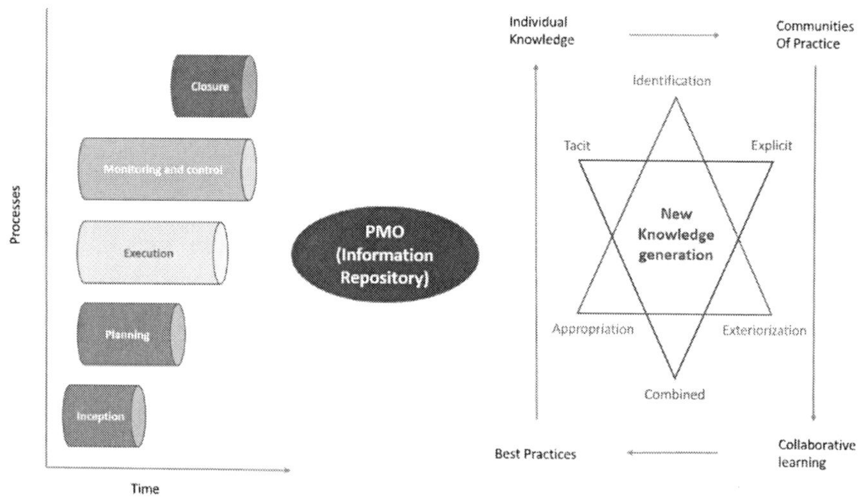

Figure 6. Articulation of the processes of the phase of Knowledge Management of the $KM+PM_{TIC}$ model.

Inception

Inception is the first process in the life cycle of the Project Management. In this moment, it is defined: the project, business needs, justification, description, main idea, specific proposal of the project, objectives, scope, quality, schedule, risks evaluation. In addition, time and costs estimation is done, taking into account human, materials and financial resources available.

Deliverables are defined at this point too, and so is reflected in the Act of Constitution of the project. This process is essential to the success of the project, because poorly planned objectives will lead to failure, even when the management is adequate.

During each of the subsequent phases of the project, verifications should be made with regard to the scope of the project for that phase. The repetition of the process of initiation in each phase allows to stop the project at any time, if it ceased to be viable for the objectives of the Organization.

Planning

Planning is the phase where the planning of all activities required to carry out the project, considering the priorities, the resources required and schedules to run each of the tasks (and their functionalities), is done. Planning refers to the identification of activities, definition of milestones and deliverables of the project and even mitigation of potential risks. Objectives, schedule and budget are defined. Intelligent planning is key to PM success. However, as new information is obtained along the project, new dependencies, requirements, risks, opportunities, assumptions and restrictions will be identified and resolved.

Execution

Execution refers to the implementation of the project and it consists of putting in practice the planning carried out previously. During the execution of the project, emphasis should be placed on communication to make decisions as quickly as possible in case that problems or inconveniences arise. In addition, regular meetings shall be organized to manage the team, i.e., to regularly discuss progress and determine the following priorities. Right now, it is the moment to evaluate the resources that are required to develop the processes planned.

Monitoring and Control

The purpose of monitoring and control activities is to ensure that the objectives are achieved in time and at the planned level of quality, by measuring the performance of the results, with the aim of taking corrective actions. This is done through the comparison between the planning carried out and the values incurred. The reports of monitoring and control must be provided in a timely manner, without delay, with the purpose of taking corrective action before it is too late. To do this, there are some recommended methods: value livestock or value of work performed method, performance measurement method and method of the milestones of payment or measurement of technical performance, among others.

Closure

According to González, Sanjuan, Pelayo & Espada (2011) the closure process of the project involves all activities through all groups of management processes of projects, to formally complete the project. In this phase they become especially important the activities related to: meet the completion criteria, transfer the products and collect the records of the project. The project manager will review all the information coming from the termination of previous phases to ensure that all work is complete and that the project has reached its objectives and goals.

This process assumes the completion of the project, all project has a temporary existence, and it ends when all stated work is reached. Closure is the final stage of a project in which revision is done, to carry out the relevant assessments on what was previously planned and what was executed, as well as their results, in consideration to the attainment of the stated objectives.

Final testing takes place with the aim to correct and /or verify the solution. A termination document should be written to describe how the project was carried out, the problems detected, the methodology used, the type of Organization in which it was executed, experience gained, and most important, the conclusions reached once the project is completed.

During the closure of the project or a phase, any of the following can occur (Project Management Institute, 2013):

- To obtain the acceptance from the customer or sponsor to formally close the project or phase.
- To perform a review after the closure of the project or the completion of a phase.
- To have recorded the impacts of the adaptation to a process.
- To have documented the lessons learned.
- To have the appropriate updates to the assets of the processes of the organization applied.
- To file all relevant documents in the repository of information.
- To close the procurement activities and ensure the end of all the relevant agreements.
- To perform the evaluation of the members of the group and release the project's resources.

- To have documented the lessons learned to use them as historical data.

Figure 7 shows the articulation of the processes of Knowledge Management, related to the project management phase.

The processes of PM phase flow into time interacting with the wisdom of knowledge that gives effect to the sextuple restraining of project development (scope, time, cost, quality, risks and resources), giving as a result, the transformation of the individual knowledge into organizational knowledge through best practices in the management integration, stakeholders management, communications and changes management, fundamental processes to reach the objectives and successful completion of the project.

The model shows how facilitators of KM and best practices of the PM can make a transfer of knowledge that ensures the collaboration from an individual atmosphere of management to one organizational that strengthens the leadership, skills and knowledge of both, the manager and the collaborator team, and therefore the organization (Figure 8).

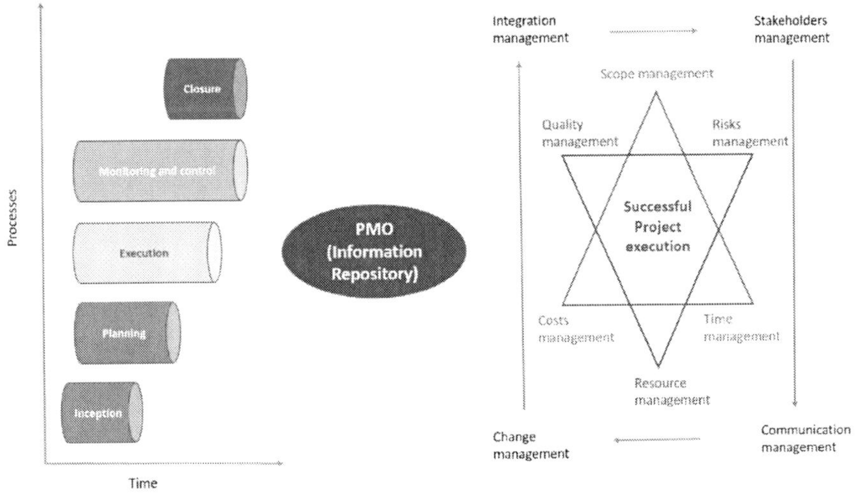

Figure 7. Articulation of the processes of the phase of Project Management of KM+PM_{TIC} Model.

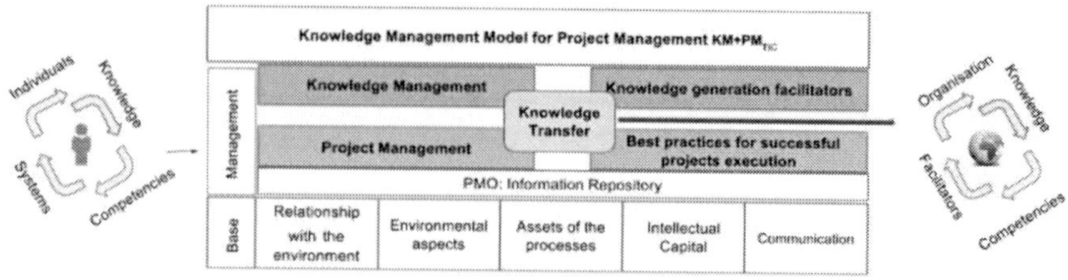

Figure 8. Result of the articulation of the processes of the phases of KM and PM of the KM+PM$_{TIC}$ Model.

Figure 9 shows the articulation of the processes of the KM phase and the processes of the phase of PM with the structure of the Project Management Office PMO, which supports the transfer of knowledge relying on the processes of knowledge that, through facilitators moving in a virtuous cycle, foster innovation and continuous improvement in the development of successful projects. In Figure 9 the consolidation of the development of the model KM+PM$_{TIC}$ is shown.

Likewise, the model shown in Figure 10 shows that the purpose is to sustain a clear and practical deployment of all the factors that affect the learning capacity of an organization, as well as the desired outcomes of learning, through the interaction of the elements defined in the phases of the management layer, which is presented as a complex system of relationships that are moved in all senses.

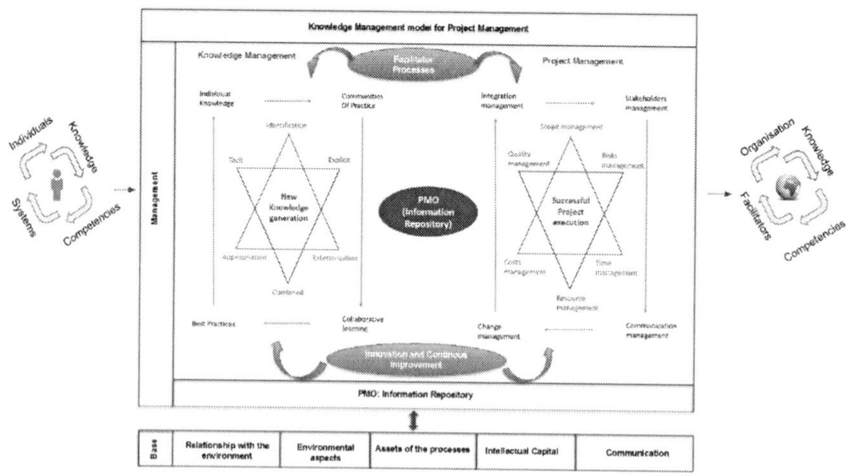

Figure 9. Knowledge Management Model for Project Management based on KM+PM$_{TIC}$.

of all its stakeholders, increasing the odds of success providing added value to the organization, to the projec of course to the project manager, who will appropriate all future transfer in the practices of projects.

sal of added value of the project management model goes ov cate time and effort to manage projects in advance, within th and achieving the objectives related to the deliverables with the acted, surpassing the expectations of the customer and uccess during the project's life cycle, as with the use and of facilitators supported by information and communication hese can be fulfilled. The benefits of the presented model are:

lete projects faster and at a lower cost: one of the greatest its of using the model is the value of reuse, as a virtuous cycle of ledge, management and improvement, which articulates the sses, procedures, forms, formats and templates, among others, ıps with minor modifications), in both, current and future :ts. The result of this is a reduced time to start projects, a lower ng curve for the members of the team, as well as time saving, ving to reinvent processes or formats from zero in each project.

time and costs with an early scope management: A large number jects have difficulties in managing the scope, resulting in efforts dditional costs. By articulating, appropriating and transferring practices in the processes management, it will be possible to the objectives to manage the scope effectively.

ons enhanced through better planning: A large number of :ts experience problems due to the gaps between the tations of the customer and the deliverables of the project. The f the model is appropriate as it strengthens the planning of the :t, which gives the team and the sponsors, the opportunity to e that there are clear arrangements between the specifications of eliverables and the needs of the customer.

er promptness for solving problems: Some teams spend a lot of and energy to face problems because they do not know how to solving them. Having an early incidents management process, to ensure that the problems will be resolved as quickly as ole.

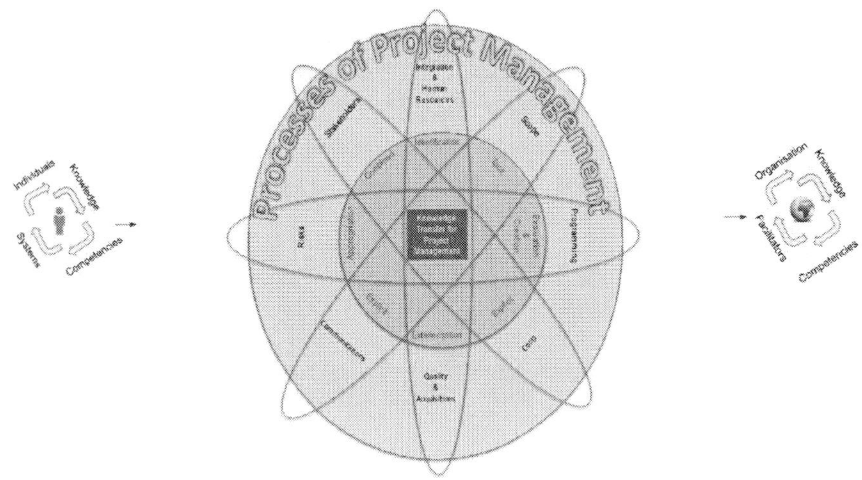

Figure 10. Interrelation of the processes of the Phases of KM and PM of the model KM+PM_{TIC}.

The model is also presented in the form of processes, in order to show clearly the triggers or entries, the stakeholders involved in the generation of new knowledge and the outputs (Figure 11).

Within this presentation, the model includes three major processes: Strategic or Managerial, Tactical or Operational and those related to Support, articulated in such a way that in each and every one of them the generation of new knowledge is permeated. The proposed model considers necessary to break barriers on the changes at the individual and organizational levels to encourage the creation of intelligent organizations. Many changes will affect the business strategy, organizational structure and the organizational culture.

In the same way, the model involves ICT that are presented as a useful and necessary tool to facilitate the communication and relationships between them. For this reason the model allows us to affirm that the KM creates a learning and collaboration environment supported by technology and decreases the organizational problems when transferring the knowledge through the use of the facilitators of KM. This is why it is fair to say that the projects that manage to align their processes of PM and KM, integrating the team with the virtual tools of collaboration, are in the right path to the natural flow of new knowledge within the organization.

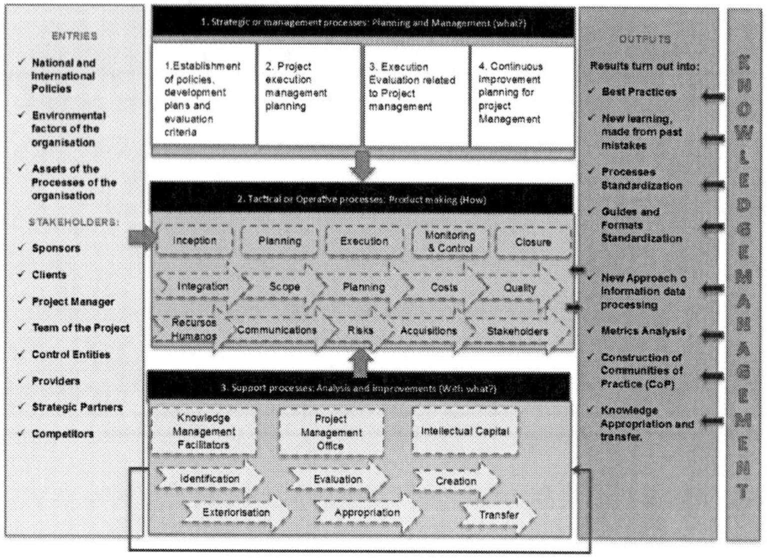

Figure 11. KM Model for PM supported in ICT - KM+PM$_{TIC}$.

Key Factors for the Success of Knowledge Management

The factors to determine the success of a process of KM are many and varied; Davenport cited in (Rodríguez, 2005), identifies nine (and interrelated as possible) key factors decisive for success:

- Knowledge oriented Culture.
- Technical and institutional infrastructure.
- Managers Backup.
- Economic value or market value bond.
- Process orientation.
- Clarity of objective and language.
- Motivation Practices.
- Structure of knowledge.
- Multiple channels for the transfer of knowledge.

Key Factors for the Success in Pro

Below, we enumerate requirements a with success:

- Strong awareness and training to
- Objectives: These are key to the p
- Scope Statement.
- Project Charter.
- Effective planning.
- WBS (Work Breakdown Structure
- Risk Management.
- Efficient system to monitor, asses
- Properly defined deadlines.
- Strong and structured Contracts.
- Implementation of the projects in
- Appropriate Quality management
- Integration and alignment betwee
- Compliance of milestones or deliv
- Establishment of a communicatio
- Requirements for contractors and
- Budgets.
- Management Changes.
- Measurement, control and improv
- Intense participation of the membe
- Effective planning of contingencie
- Committed leadership.
- Team.

CONCLUS

The Knowledge Management model fo ICT, KM+PM$_{TIC}$, supports the use of ex practices to provide an outline of act, ba (Project Management Institute, 2013). standards and techniques for managing

teamwork a consequentl sponsors an knowledge f

The pro this idea: de allotted budg quality con maintaining appropriatio technologies

- Con bene kno proc (per proj lear not
- Sav of p and best defi
- Solu proj exp use proj ensu the
- Gre time star help poss

- Anticipate risks and mitigate the inevitable, before they materialize into bigger problems: All Project Management processes include processes to identify and manage risks. Risk management provides the identification of potential problems and the corresponding mitigation plan, with the purpose of supporting those activity quotas, before they arise.
- Manage communication processes and expectations with customers, team members and stakeholders, more effectively: A large part of the problems with projects is derived by the deficiency of effective channels of communication. These are avoided with a communication system articulated among project stakeholders, to ensure full knowledge of problematic situations, which may delay it or make it more costly. It is important to avoid incidents, with formal or informal effective communications.
- Create higher-quality products and services: Quality Management processes will help the team to understand the needs of the customer in terms of quality. Once these are defined, the team can implement controls and techniques that will make it possible to meet the expectations of the customer.
- Improve the financial management achieving greater benefits to the sponsor and therefore, for the organization: This is the result of a better definition of the project, better estimations, a formal budgeting process and a better monitoring of actual costs versus the budget. This entire rigor means better forecasting and financial control. There is more information available that will enable the Organization to stop an "evil" project with greater opportunity.
- Stop "bad" projects with greater promptness (Díaz, Medina, González & Liberona, 2014): A project may have begun with a clear cost-benefit ratio. However, if the project takes too long and has exceeded its budget, it can reach the limit at which business is no longer profitable. The effective project management allows timely identification of this type of situation, so that decisions can be taken to refocus the scope or to cancel the project.
- Achieve greater monitoring and control to the management indicators, feeding back on time, decision making based on real facts: One of the most sophisticated aspects of methodologies is that they provide a guide to make it easier to collect metrics (indicators - measures). The indicators provide information that helps determine how effective and

efficient the performance of the team is and the quality levels of deliverables.

- Work environment and teamwork improvement: If the projects are successful, you will find additional benefits associated to the team. Customers have greater involvement; the team will have a better sense of ownership and behave with greater trust and professionalism. People who work in projects with problems tend to be unhappy. On the other hand, the people who work in successful projects tend to feel more motivated and satisfied with their work and with themselves.

Moreover, the implementation of projects that use guidelines, methodologies or models of management, can expect in addition, the following benefits:

- Increased productivity due to the use of standards across the macro processes defined in this research (start, planning, execution, monitoring and control, and close).
- Increased profitability of the project due to the use of best practices in the processes of planning and control of the resources in the project.
- Reduction of the resources that are applied to projects without added value.
- Standardization of practices, processes, procedures, forms, formats, guides and checklists, among others, in all areas of the project and thus, the Organization.
- Standardized management systems to follow, in real time, companies, industrial segments, suppliers, customers, sponsors and regulatory bodies (stakeholders).
- Development of activities geared to processes with knowledge needed to manage the lifecycle of any project, program and portfolio.

REFERENCES

ADECEC. (2013). Guide, Storytelling and Branded Content [Online]. Retrieved from http://www.adecec.com/pdf/la_guia_de_storytelling_y_branded_content_de_adecec.pdf.

Bernandez, M. L. (2007). Diseño Producción Implementación de E-learning. Bloomington, Indiana: AuthorHouse.

Bonilla Gutierrez, C. (1988). La comunicación, función básica de las Relaciones Públicas. México: Trillas.

Cerpa, N., & Verner, J. M. (2009). Why Did Your Project Fail?. *Communications of the ACM, 52*(12), 130-134.

De Long, D. W. & Fahey, L. (2000). Diagnosing cultural barriers to knowledge management. *Academy of management, 14*(4), 113-127.

Díaz, F. N., Medina, V., & Joyanes, L. (2013). Technological Tools Virtual Collaborative to Support Knowledge Management in Project Management. In L. Uden, F. Herrera, J. Bajo & J. M. Corchado (Eds.), *7th International Conference on Knowledge Management in Organizations: Service and Cloud Computing: Vol. 172. Advances in Intelligent Systems and Computing* (pp. 163-174). Springer-Verlag Berlin Heidelberg.

Díaz, F. N., Medina, V. & González, R. (2014). ICT as a means of generating knowledge for Project Management. In L. Uden, L. S. L. Wang, J. M. Corchado, H.-C. Yang, I.-H. Ting. (Eds.), *The 8th International Conference on Knowledge Management in Organizations: Social and Big Data Computing for Knowledge Management. Springer Proceedings in Complexity* (pp. 617-629). Springer+Business Media Dordrecht.

Díaz, F. N., Medina, V., González, R. & Liberona, D. (2014). Knowledge management, innovation and efficiency of service enterprises through ICTs appropriation and usage. In L. Uden, D. F. Oshee, I.-H. Ting, & D. Liberona (Eds.), *The 9th International Conference on Knowledge Management in Organizations: Vol. 185. Lecture Notes in Business Information Processing* (pp. 300-310). Springer International Publishing Switzerland.

Díaz, F. N., González, R., & Medina, V. (2015). Analysis and Improvement of the Management of IT Projects. *IEEE Latin America Transactions, 13*(7), 2366-2371.

Díaz, F. N., Medina, V. & González, R. (2015). Knowledge Management Model for Project Management. In L. Uden, M. Hericko, & I.-H. Ting (Eds.), *Knowledge Management in Organizations: Vol. 224. Lecture Notes in Business Infromation Processing* (pp. 235–247). Springer International Publishing Switzerland.

Esterkin, J. (2008). What is expert judgment? Best projects [Online]. Retrieved from: http://iaap.wordpress.com/2008/02/22/que-es-el-juicio-de-expertos/.

Galunic, D. & Rodan, S. (2007). Resource recombination's in the firms: knowledge structures and the potential for Schumpeterian innovation. *Strategic Management Journal, 19*, 1193-1201.

González, R., Sanjuán, O., Pelayo, B., & Espada, J. (2011). Management and Project Management. College of Engineering and Architecture. Pontifical University of Salamanca. Madrid Spain.

Gudi, A. & Becerra-Fernandez, I. (2006). Role of knowledge management in project management of complex systems organizations. In *NASA Knowledge Management and Successful Mission Operations Conference 2006*. NASA's Base of Knowledge.

Hillman, P. (2010). According to Gartner, 75 percent of all IT projects fail. Yours don't? Are you sure? [Online]. Retrieved from: www.cdh.com.

Humaidi, N., & Asarani, N. (2012). Investigation on project management performance Using Knowledge project management Performance Assessment model: A pilot study. *International Journal of Innovation, Management and Technology, 3*(6), 769-772.

Jaramillo, D. (2012). Diseño de un sistema de gestión del conocimiento para la dirección de servicios y recursos de información de la Universidad ICESI (Doctoral Thesis). Universidad del Valle, Santiago de Cali.

Lozano, M. (2010). National Strategy for Social Appropriation of Science Technology and Innovation [Online]. Retrieved from: http://www.eafit.edu.co/foro-taller-ascti/index.html.

Medellín, V. (2006). Project management. Bogota: Editorial Universidad del Rosario Center.

Medina, V. (2004). Organizational Model and Knowledge Management Technological University: Application in the University Francisco José de Caldas. Doctoral Thesis. School of Computing. Pontifical University of Salamanca. Spain.

Morreal, R. (2012). The Top Reasons Projects are Unsuccessful. *PM World Journal, 1*(5) [Online]. Retrieved from: http://articulospm.files.wordpress.com/2013/11/the-top-reasons-projects-are-unsuccessful.pdf.

Myriam, B. A. (2007). Building Collaborative Learning Object. In *II Latin American Conference on Learning Objects LACLO 2007*. Latin America.

Nonaka, I. & Konno, N. (1998). The concept of Ba: Building a foundation for knowledge creation. *California Management Review, 40*(3), 40-54.

Nonaka I. & Takeuchi, H. (1999). *The creative organization of knowledge*. México D. F: Editorial Oxford Press.

Nonaka, I., Reinmoeller, P., & Senoo, D. (1998): The art of knowledge: systems to capitalize on market knowledge. *European Management Journal, 16*(6), 673-684.

Nonaka, I., Toyama, R., & Konno, N. (2001). SECI, Ba and Leadership: a Unified Model of Dynamic Knowledge Creation. *Long Range Planning, 33*(1), 5-34.

Project Management Institute (2013). *A Guide to the Project Management Body of Knowledge PMBOK Guide* (Fifth Edition). Pensylvania: Project Management Institute.

Rodríguez, J. (2005). *IT project management: methods, tools and cases.* Catalunya: UOC Editorial.

Sanchez, M. L. (2005). Desarrollo de la comunicación externa en la empresa. Electronic bulletin AEG - Articles of interest. Graduates Alumni Association PUCP [Online]. Retrieved from: http://aeg.pucp.edu.pe/boletinaeg/articulosinteres/32/comunicaciones_sanchez.pdf.

Smith, A., Bieg, D. & Cabrey, T. (2014). PMI's Pulse of the Profession® In-Depth Report: Requirements Management. A Core Competency for Project and Program Success. Project Management Institute [Online]. Retrieved from: http://www.pmi.org/~/media/PDF/Knowledge%20 Center/PMI-Pulse-Requirements-Management-In-Depth-Report.ashx.

Standish Group. (2015). *Chaos Report 2015.* The Standish Group.

Stewart, T. A. (1991). Brainpower Intellectual capital is becoming corporate America's most valuable asset and can be its sharpest competitive weapon. The challenge is to find what you have -- and use it. Fortune Magazine [Online]. Retrieved from: http://archive.fortune.com/magazines/fortune/fortune_archive/1991/06/03/75096/index.htm.

Santamaría, F. (2007). Herramientas colaborativas para la enseñanza usando tecnologías web: weblogs, wikis, web 2.0 [Online]. Retrieved from: http://tadiguea.wordpress.com/2007/12/14/weblogs-wiki-redes-socialesweb-20/.

Tan, S. (2011). How to Increase Your IT Project Success Rate [Online]. Retrieved from: https://www.gartner.com/doc/1531025/increase-it-project -success-rate.

The Standish Group International, Incorporated. (2013). Chaos Manifesto 2013 Think Big, Act Small [Online]. Retrieved from: http://www.versionone.com/assets/img/files/CHAOSManifesto2013.pdf.

Valhondo, D. (2003). Gestión del conocimiento: del mito a la realidad. Madrid: Díaz de Santos.

Yeong, A. & Lim, T. (2010). Integrating knowledge management with project management for project success. *Journal of Project, Program & Portfolio Management, 1*(2), 8-19.

Yin, R. K. (1989). The Case Study Crisis: Some Answers. *Administrative Science Quaterly, 26*(1), 58-65.

Reviewed by: Dr. Alberto Magreñán, Real Spanish Mathematical Society, Spain. Dr. Carlos Enrique Montenegro Marín, Francisco José de Caldas District University, Colombia.

In: Construction Projects
Editor: Kimberly Hall
ISBN: 978-1-53610-742-5

Chapter 3

LEVERAGING THE BLOCKCHAIN TECHNOLOGY TO IMPROVE CONSTRUCTION VALUE DELIVERY: THE OPPORTUNITIES, BENEFITS AND CHALLENGES

***Oliver Barima**, PhD**
George Brown College, Toronto, Canada

ABSTRACT

Bitcoin, the first cryptocurrency in the world has been in use since 2009. Blockchain is a technology that underlies the use of bitcoins. In recent times, blockchain technology is being touted as a promising tool which can transform many business operations. Indeed, it has been described as the next big thing after the emergence of the internet. This study examines the blockchain technology with the view to informing the construction industry on the potential of this technology to enhance value delivery. The study explores the potential uses, benefits, challenges and implications of the blockchain technology for the construction industry. Additionally, it provides future research areas for the leverage of the blockchain technology in project delivery and other construction-focused activities.

---

* Corresponding author: 1240 Eglinton Ave West RPO 37066, Miss.; ON, Canada, Email: obarima7@gmail.com.

Keywords: bitcoins, blockchain technology, value delivery, construction industry, construction project delivery

INTRODUCTION

In recent times, innovation has become a critical variable for value delivery, competitive strength and organizational survival (e.g., Barima, 2008a). Hence, it is not surprising in contemporary times to find organizations (sometimes backed by sovereign nations) seeking to leverage their innovative skills to outperform the competition, increase their market share, provide better value to their stakeholders and enhance organizational survival in the long term.

The business environment has been turbulent and very competitive in recent times (e.g., Barima, 2008a). Additionally, significant leaps in product and process innovations have occurred in this contemporary period (e.g., Barima, 2008a). In certain cases these advances have unleashed enormous disruptions to extant ways of business operations. For example, digital technology and allied tools have in recent times completely transformed several business operations and industries across the world.

However, the effect of technology as a source of competitive advantage moves through phases. As extant technologies become base technologies in any industry (for the critical core of users), organizations which previously used to dwell on such technologies as a source of competitive advantage over their rivals will continuously see their superiority disappear. The big challenge for many organizations is the ability to continuously innovate and find new competitive niches to strengthen value delivery and competitive advantage.

Rationally, many organizations are increasingly becoming restless as they seek for the next big thing. To such companies being aware of trends in technology/innovation from within and without of their operational zones is one critical way to enhance innovation, corporate vision and also build proper basis for strategic and tactical superiority. Asking the key question of what is the next big thing is a rational inquiry for organizations to demand and explore from both within and without of their corporate borders. However, this has never been an easy question to answer.

The scenarios sketched above are not different for the construction industry. Process and product innovations have been at the forefront of certain progressive organizations in the industry (e.g., Barima, 2008a). Within such organizations, the constant quest to seek the next big thing to enhance their value delivery mechanisms will not be any surprising revelation. In terms of such persistent inquest, one recent relevant technology touted to have a potentially significant promise in the business realms is the blockchain technology.

The blockchain technology is the underlying technology behind the use of Bitcoins used in business transactions (BBC, 2016; Peck, 2015; Treasury, 2015; Zoe, 2016). This technology has fascinated the world in recent times (e.g., BBC, 2016; Peck, 2015; Treasury, 2015) and diverse people from sectors such as government, international public organizations, financial sectors (e.g., bankers), Information Technology(IT), insurance, etc. have been riveted by the potential of the technology (BBC, 2016; Cellan-Jones, 2016; Higgins, 2016; Peck, 2015; Peck, 2016; Postscapes, 2016; Treasury, 2015). Indeed, certain people have described the use of the blockchain technology as the next big thing after the significantly influential emergence of the internet (BBC, 2016; Peck, 2015; Peck, 2016; Postscapes, 2016; Treasury, 2015).

The blockchain technology has been predicted to have the potential to affect business transactions and this certainly includes the construction industry (BBC, 2016; Peck, 2015; Peck, 2016; Postscapes, 2016; Treasury, 2015). This latter premise raises a number of questions for actors in the construction industry including the following. What is the blockchain technology? What are the implications of this recent interest in the blockchain technology for the construction industry? What are the potential uses, opportunities and challenges of the blockchain technology in the construction industry? What are some of the construction industry focused future areas of research to enhance the use of this technology?

The aims of this paper are as follows. First, the paper examines the block chain technology from a generic perspective with the view to informing the construction industry on the potential of this technology to enhance value delivery. Second, the study seeks to explore the potential opportunities, benefits, and challenges of this technology from a construction industry perspective. Third, the study distills implications for practice and provides recommendations for future research in the leverage of the blockchain technology in the construction industry.

This paper is organized as follows. First, the study discusses Bitcoins, blockchain and allied technologies to enhance understanding of the topic. Second, the study explores the potential opportunities, challenges and benefits in the use of the technology in the construction industry. Third, the study provides implications for practice and also areas for further research on the topic. Finally, the paper provides a synthesis and conclusion to the discussion.

BITCOINS, BLOCHCHAINS, ALLIED TECHNOLOGIES AND EMERGING TRENDS

Bitcoins and Blockchain Technology

Blockchain is a system for recording data (BBC, 2016; Peck, 2015; Peck, 2016; Postscapes, 2016; Treasury, 2015; Zoe, 2016). These data can be in the form of digital registers of agreements, transactions, contracts and also things which require independent recording and verification of their occurrence (BBC, 2016; Peck, 2015). Though this idea has existed for decades (and hence it is not entirely new), the invention of the Bitcoin enhanced the importance of the concept (BBC, 2016).

Blockchain is the technology that underlies the Bitcoin technology (Peck, 2015; Zoe, 2016). Bitcoin technology is a recent innovation that made its debut on the market in 2009 (Postscapes, 2016; Treasury, 2015). Bitcoins are the results of complex mathematical computations by computers, and in the course of these operations they consume very significant computing power and electricity (BBC, 2016). As the world's first cryptocurrency, Bitcoins were intended to facilitate digital payments and transactions (Postscapes, 2016; Treasury, 2015; Zoe, 2016).

Cryptocurrency is a type of virtual money that can be conveyed in a secure and anonymous manner across a wide peer-to-peer distributed network (Postscapes, 2016). Thus, it is a form of an inconvertible fiat. The Bitcoin blockchain technology primarily acts as a ledger/register which registers all legitimate transactions which occur on the network within a specific time frame (Postscapes, 2016; Treasury, 2015). And all the nodes in a network collaboratively maintain this ledger to make the Bitcoin blockchain process entirely transparent, auditable and automatic based on consensus (BBC, 2016; Peck, 2015; Postscapes, 2016; Treasury, 2015). The Bitcoin network automatically guarantees a fair and accurate register of who owns specific Bitcoins in specific timeframes, and hence this removes the need to entrust third parties to do so (Peck, 2015; Postscapes, 2016; Zoe, 2016).

A typical Bitcoin transaction goes through the following processes. An encrypted transactional record is sent to all nodes in a Bitcoin network every time two parties exchange Bitcoins (BBC, 2016; Peck, 2015; Postscapes, 2016). The encryption process is called 'hashing' and this is done by many diverse computers (BBC, 2016). Digital records are combined into 'blocks' and then they are cryptographically and chronologically turned into chains via complex algorithms (BBC, 2016; Peck, 2015; Postscapes, 2016; Zoe, 2016). Whenever consensus is reached then each block obtains a unique digital signature (BBC, 2016). The computing nodes conduct complex cryptographic computations on the extant recorded data to verify each transaction (BBC, 2016; Postscapes, 2016). Then they give notice to all the nodes anytime a new 'block' of legitimate transaction is confirmed (BBC, 2016; Postscapes, 2016; Treasury, 2015). If the bulk of the nodes pass this block then every node adds this to the ledger/register (BBC, 2016; Postscapes, 2016; Treasury, 2015). The updated version of the ledger/register is then used as a new basis for encrypting and authenticating prospective transactions (Peck, 2015; Postscapes, 2016; Treasury, 2015). The details of the transactions are not stored, but rather only the 'hash' of the transaction and also the fact that they happened (BBC, 2016). Additionally, the blockchain technology enables the creation of assets and the transfer of digital information without copying (Zoe, 2016).

Whenever a ledger is updated it is difficult to alter or tamper with it (BBC, 2016; Peck, 2015). The continuous extension and collective verification of blockchains make them potentially difficult for fraud (apart from any extant flaws in the basic structure) since blockchains register accurate tracks of transactions/exchanges (BBC, 2016; Peck, 2015; Postscapes, 2016; Treasury, 2015). The distributed nature of blockchains makes it difficult for hackers to hack it, since to do so they will require very significant computing power and also simultaneous access to all the nodes to be successful(BBC, 2016; Peck, 2015).

Since the emergence of Bitcoin technology about 7 years ago, the technology has been used to facilitate all manner of payments. Bitcoins were once promoted as a global currency free from any government controls (Zoe, 2016). So far, Bitcoin transactions have included legal purchases of items via the web, to the leverage of the technology by criminals to extort money from innocent parties (see e.g., Campbell, 2016; CBC, 2016). The negative perceptions of the technology are in part due to the activities of criminals whose nefarious activities with this medium have given the word 'Bitcoins' a bad press in recent times(see e.g., Campbell, 2016 ; CBC, 2016).

Perhaps, fear is not a good way to highlight the potential benefits of an emerging technology in the construction industry. In spite of the recent bad press, the use of Bitcoins as a payment system has opened up knowledge into other potential uses of the technology behind the Bitcoins (BBC, 2016; Peck, 2015). This is a significant revelation since initially the potential applied uses of the blockchain technology were not touted.

Alt-Chains, Sidechains

Altchains and sidechains are important developments which have spawned out of the Bitcoin blockchain technology. The following sections describe these allied technologies (BBC, 2016; Peck, 2015; Postscapes, 2016):

- Other cryptocurrencies such as '*Altcoins*' have emerged since the debut of Bitcoins in 2009. These are often based on similar structures as Bitcoins and they mainly represent bifurcations of the Bitcoin software. Apart from monetary transactions other non-restrictive uses of blockchains were quickly realized after the introduction of Bitcoins in 2009. Indeed, it was discovered that they could be used to monitor and confirm all kinds of digital exchanges. Examples of the other uses include 'Namecoin' and 'Datacoin' used for keeping/controlling records of names and data in their respective blockchains.
- **Altchains** have proliferated, but they face a number of challenges. The security/reliability of blockchain networks are linked to the number of constituent nodes. And Bitcoin has a huge lead in terms of computational power and so this makes it difficult for new cryptocurrencies to find their grounding. Also, altchains duplicate most of the codes and functions of other blockchains without direct integration with them. This is partly due to the reluctance of Bitcoin developers to change the software (given that there is much at stake now).
- **Sidechains:** have been proposed to deal with these latter problems. Sidechain technologies enable the creation of software which iterate on root blockchains. Additionally, this allows assets to be transferred directly between parent chains and sidechains as well as between sidechains. This feat opens the way for the use of sidechains to exchange names, data, and other assets within the Bitcoin network without changing the core functions of Bitcoin.

Indeed, the spawning of various systems of blockchains has also created the need for inter-ledger operators/entities that enable the exchange of digital assets between blockchains (e.g., Zoe, 2016). Additionally, assets such as Bitcoins or Namecoins cannot be directly transferred between blockchains (but rather third parties provide this service) (e.g., Postscapes, 2016; Zoe, 2016).

Emerging Consortiums, Blockchain Middleware and Cryplets

In recent times several blockchain consortiums are emerging to test the technology. These include banks (e.g., consortium R3CEV, for more than 40 banks globally), trading organizations and certain countries (e.g., The USA, United Kingdom, Dubai's Global Blockchain Council (GBC)) (BBC, 2016; Castillo, 2016; Cellan-Jones, 2016).

Thus, this encouraging drive is not only emerging in the private sector alone, but also the public sectors in certain places. Notably, they include the UK government, US government, and international entities such as the military alliance, NATO, and the Commonwealth Secretariat. For example, recently the UK government quietly launched a trial of the blockchain technology in relation to the payment of benefits in a bid to help claimants manage their money (Cellan-Jones, 2016). This is in line with the recommendations of the Chief Scientific Advisor for the government to back trials of the technology whilst issues of security and privacy implications are carefully considered (Cellan-Jones, 2016).

Also, the US government via the Department of Homeland Security (DHS) recently gave a grant of $600,000 to six companies executing work on blockchain applications for the government (Higgins, 2016). Organizations such as NATO and the Commonwealth Secretariat, the executive arm of the 53-member Commonwealth of Nations, have also shown recent interest (e.g., Higgins, 2016). For example, NATO announced a recent technology innovation contest in May this year (2016), whilst the Commonwealth Secretariat also unveiled a 'messaging app' designed with applications for law enforcement which apply the technology to establish identity (Higgins, 2016). These moves are commendable, since almost always governments play significant roles in the development of infrastructure to support technology development as was in the case of the internet (Leising, 2016).

Other noteworthy programs include Ethereum and the efforts of Microsoft. Ethereum provides user-friendly platforms that can be used to deploy software on a blockchain network (Peck, 2016). Recently Microsoft's director of business development and strategy, Marley Gray announced Project Bletchley, a vision to provide Blockchain as a Service (BaaS) which is flexible and open for all platforms, partners and customers. As part of this service the company has two new concepts called blockchain middleware and cryptlets (Castillo, 2016). The middleware is intended to deliver key services in the cloud such as identity and operations management, data analytics and machine learning (Castillo, 2016). These services can help users get the benefits of increased security and immutable record keeping via the blockchain technology and also deliver business intelligent reports needed by regulators, etc. (Castillo, 2016). Cryplets on the other hand will aid users obtain extra information such as dates/times needed to execute contracts (Castillo, 2016). Also it will enable secure interoperability between customer technologies, ecosystem middleware and Microsoft Azure (Castillo, 2016).

On the hand, according to Allison (2016), big technology companies such as Amazon, Google, Facebook and Apple have so far been visibly silent on the blockchain technology. This silence has made people such as Vitalik Buterin, an inventor of Ethereum, to question if the 'blockchain blindspot' around such big companies (i.e., Google, Facebook and Apple) is really an indication of the threat which the technology poses to these types of companies (Allison, 2016). It is better, however, not to speculate based on such a premise, since perhaps these latter innovation-oriented companies may have specific strategies which they may reveal in due course.

Apart from the descriptions above, the types of access to the blockchain technology vary. Access can be via open or closed sources. An open access (i.e., to all parties) is called 'unpermissioned' access and a restricted access (i.e., to only a selected group of people) is called 'permissioned' access (BBC, 2016; Zoe, 2016). The latter type of access which ensures that only trusted parties have access to the technology is, of course, an approach that is favored by banks (e.g., Zoe, 2016). Over time a number of hybrid types of accesses to the use of blockchain technologies are likely to emerge widely across industries.

Opportunities and Benefits in the Use of the Technology in the Construction Industry

The Construction Industry in Context

The discussion in this section will be enhanced by first placing the nature of the construction industry in context. Construction industry forms a significant aspect of every economy. In the year 2015, the industry accounted for about $8.5 trillion dollars globally in terms of spending (Timetric, 2016). The industry often provides products which are used by other sectors of the economy to deliver products and services in national economies.

From a generic perspective, the construction industry is a fragmented and a low margin industry. Works are often procured in the industry via competition, negotiation, or hybrids of these latter methods. Governments are often the largest clients of the construction industry (Barima, 2008b). And usually, based on international agreements in many places, procurement for most public works are done on a competitive basis. The industry is often slow to adopt technology and comparatively the industry often lags behind other industries in terms of innovation and productivity improvements (e.g., Barima, 2008a; Government of Canada, 2016).

A useful way to understand the usefulness of technology adoption in the industry is via comprehension of the typical asset lifecycle of products produced in the industry. A typical asset produced in the construction industry moves through the phases of concept, development, implementation and phase-out during project delivery phase. There are many product delivery models used in the construction project delivery phase. These include systems such as the following: traditional design-bid-build; design and build; design-build-operate; design-build-operate-transfer; public-private partnerships, etc.

The post-project delivery phase is also a very important period. The operation and maintenance phase of the delivered products covers a significant aspect of the entire life of the delivered assets till disposal. Given the durability of the products delivered in the industry, the latter phase can cover many years, even centuries in certain cases, till final disposal.

The significance of the use of the blockchain technology in the construction industry can be meaningfully seen via the link between the technology, the phases of a typical asset and the role of other relevant extant and emerging tools that can be applied in tandem in the industry to provide synergistic benefits. The next section examines these issues.

Potential Uses in the Industry

Apart from the use of Bitcoin/altchain/sidechain blockchain technologies to support transactions such as payments, names and data, the use of the technology in other areas is nascent. Discussion on the use of the technology can therefore be seen from extant and aspirational uses. From a construction perspective the following as discussed below can emerge.

One key area which can benefit from the use of the blockchain technology is in the domain of construction procurement. This often entails the procurement of services, materials and also the works. Traditionally, each of the procurement processes for services, materials or the works differ, but they may also hold certain things in common. For instance, procurement for the works (given the relative monetary value of the works/number of parties involved) is often more complex than the other indicated two types of procurement. The models usually suitable for the procurement of materials will often not be fit for that usually applied for the works.

The blockchain technology can be used to monitor records/registers between devices, and devices and their users/web services in a transparent manner in each of the identified procurement types (e.g., Postscapes, 2016). For example, during the procurement of materials activities such as the monitoring of stocks, generation of requisitions, the placing of orders, expediting, transportation, delivery of orders, quality control, acceptance of the orders to the specific site/storage areas can be monitored and dealt with in a transparent and accountable manner via the blockchain technology.

The blockchain system can also be used for instant automatic electronic payment when all the prescribed requirements of the procurement agreement are met. Systems can also be in place to monitor and estimate stock levels, the rate of use of specific materials at various construction sites and then trigger bids from recognized traditional construction suppliers or alternatively source bids from similar autonomous machines (e.g., Postscapes, 2016). Additionally, data created in the indicated processes can support audits or help resolve disputes that may occur in the course of the indicated transactions. However, given the nascent nature of the technology a graduated approach may be necessary to forestall significant risks in the initial adoption of technology.

Currently, certain works are being done on procurement with the support of the blockchain technology. Traditionally, construction bid competition can be highly redundant, increase variability and also consume significant resources (Robles, 2015). Robles (2015) describes a form of smart contract application (i.e., based on blockchain) called Bid-Pool which is intended to lower the costs of bidding from A X B to A/B. Where A = cost of goods sold and B = number of bidders. Also reward high integrity in the bidding process. According to Robles (2015) this will eliminate bidding redundancy, reduce marginal costs of additional bidders, reduce corruption and conflict of interest due to transparency.

Governments are often the major clients of construction products and services (Barima, 2008b). Therefore the attitude and the drive of governments (around the world) to use the blockchain technology will be very significant. Currently, certain governments are anticipating the need to adopt blockchain technologies (BBC, 2016; Peck, 2016). This drive can affect the way contracts, permits, regulatory requirements, and other government services are mandated by governments to be delivered via the blockchain system. The drive and actions of governments in this direction can change many things in the way extant business transactions are executed. For example, there can be significant implications on how governments and central banks regulate and make new requirements for the payments for certain works (e.g., public) done in the near future in the construction industry.

Currently, there is a growing interest by central bankers and certain influential international financial organizations in the blockchain technology. At a recent international conference in Washington, USA, the head of a blockchain technology start-up company was provided the opportunity to address central bankers from 90 countries including the US Federal Reserve as well as entities such as the World Bank, International Monetary Fund, and Bank for International Settlements on the blockchain technology (Leising, 2016). The bankers were advised to seize the golden opportunity to develop central bank digital currency (Leising, 2016).

Also, all things being equal, government revenue may increase as transactions become transparent for tax and regulatory purposes (e.g., BBC, 2016). Thus if the same system is not used negatively to avoid paying taxes. In the case of better tax revenues, this can have the indirect effect to improve government spending on capital projects. On the other negative side this can also negatively affect revenues.

One critical area in property development relates to registration of land titles. This may be taken for granted in certain places, but in the developing world this may be non-existent in certain cases (e.g., Roszak, 2016). The registration of land in a transparent and auditable manner via the blockchain technology can deal with issues of fraud, and give cost-effective access to many disadvantaged people (e.g., Roszak, 2016). The entrenched database can also be used by various owners in different parts of the world to instantly check titles to avoid potential problems with property ownership rights and construction project delivery. For example, this knowledge can avoid potential court injunctions which are sometimes issued (during land title disputes) to halt the progress of construction projects.

Certain blockchain technology-based programs can potentially be used to manage smart interconnected machines via the internet of things (e.g., Postscape 2016). This can provide transparent opportunity for learning, conduction of machine-to-machine transactions and also the assignment of responsibilities when liabilities occur. Especially, this opportunity will be very useful during the facilities management phases of constructed products.

Another area that may be potentially affected by the pervasive use of the blockchain technology in the construction industry is the use of financial institutions/intermediaries by construction actors. Currently, financial companies are the nerve in construction payments/finance, but the future role of such institutions in the pervasive use of the blockchain system over time is currently unclear. There can be a complete metamorphosis or even the elimination (in certain cases) of the role of such financial intermediaries in the industry in the future.

If extant financial institutions are able to strengthen their role, then as the technology matures, transaction fees (e.g., bank fees) are likely to be reduced due to the efficiencies and reduction in transaction costs which can be attained via the blockchain networks (e.g., BBC, 2016; Peck, 2016; Peck, 2015). Also, this can lead to faster/easier transfer of payments/money across borders with little or no government controls (BBC, 2016).

In international construction value delivery, exchange of money across borders is often a challenge and also currency fluctuations can add uncertainty and constraints in project delivery for very large projects. Via the use of the blockchain system, middlemen and many inefficient interventions (e.g., manual) can be removed from banking/financial operations (BBC, 2016; Leising, 2016). This feat can facilitate cross-border international construction projects/programs delivery and financial transfers to effect payments for goods and services in a cost effective, transparent and fast manner. Currently, the use of certain systems(e.g., Swift) may lead to days/weeks in delays in monetary transactions and this limitation locks up billions of dollars as a result, hence the blochchain technology can free up the systems and also potentially reduce threats in fraud (Leising, 2016).

In terms of security and privacy issues, there have been attempts by certain organizations to apply the blockchain technology to add security and privacy to data to deal with threats from unscrupulous people, especially in

this era of big data. For example in Estonia, a version of blockchain technology (i.e., keyless signature infrastructure) is being applied to support the government to manage and protect the data of citizens across about 1,000 online services (BBC, 2016). Also, certain people have argued that attacking the blockchain system requires large amounts of computing power (hence electricity) and significant consensus among miners to do so (Peck, 2015). These latter difficulties/challenges inherently enhance the security of the technology.

CHALLENGES IN THE LEVERAGE OF THE BLOCKCHAIN TECHNOLOGY

Generic Challenges

The blockchain technology is currently a work-in-progress and many works are being done to discover and improve its utility (BBC, 2016; Peck, 2016). At the current state of technology maturity there are many unknowns and the following are some of the concerns relating to the use of the blockchain technology.

- Any type of imaginable software can run on blockchain in its generalized form (Peck, 2016). Programs run automatically on blockchains and they can be accessible to internet users (Peck, 2016). Also they are basically autonomous and almost difficult for governments to control (Peck, 2016). The lack of proper government controls can be a significant threat as this can increase corruption and money laundering to finance criminal activities in the construction industry.
- In spite of the hype concerning the blockchain technology certain developers believe that it can be used anonymously and someone can specifically design a program to breach private information/databases and hence infringe on privacy (Peck, 2016). However, work is in progress to deal with this latter problem (Peck, 2016).
- Developers have warned of the danger in hosting certain critical things such as property title systems (by governments) on blockchains, since this is an algorithm that will not have the ability to respond to subsequent court orders not specified in the algorithm

(Peck, 2016). The autonomous nature of the software will make it difficult for society to regulate and stop certain types of actions on the blockchain (Peck, 2016). At the protocol level, there is little/nothing that can be done to ensure that people use the blockchain to only build or deliver 'good things' (Peck, 2016).

- The construction industry is often slow in adopting new technologies (e.g., Barima, 2008a). Hence, this generic attitude can be a key challenge in the adoption of the blockchain technology, and major efforts will be needed to deal with this orientation.
- Currently, knowledge in the use of the blockchain technology is embryonic, and not much data exist on the costs associated with technology adoption. Most technologies are often not initially cheap till costs gradually decrease with maturity over the years. Costs can be a key issue in the adoption of emerging technologies such as blockchains in the construction industry.
- The current state of awareness in the potential uses, opportunities and challenges in the leverage of the blockchain technology is scanty in the industry. A key challenge exists in the industry to disseminate or discuss these issues in a construction-focused manner to boost adoption.

Bad Press and Selling the Idea; Hype or Reality

The bad press about the use of Bitcoins by criminals was highlighted earlier on in this paper. For example, several organizations have suffered from the pranks of criminals who are able to hijack the interlinked computer systems of organizations via the use of ransomware to attack these systems (Campbell, 2016; CBC, 2016). Usually, these systems are taken captive when unsuspecting individuals click on infested computer links devised by the criminals. To redeem infested systems, criminals often demand payment via Bitcoins, in a manner that sometimes makes it difficult to track them. A typical case in point is the University of Calgary, AB, Canada, which recently announced the payment of $20,000 in Bitcoins to criminals in order to redeem their computing systems which were paralyzed via ransomware attacks by criminals (Campbell, 2016; CBC, 2016).

The use of the Bitcoins which is directly related to the blockchain technology can militate against the selling of the technology to certain people in the industry. The key issue in this latter narration is a concern for potential criminal activities via the proliferation of the technology. To overcome this initial bad perception, it will require significant advances/assurances to deal with such perceived threats.

This is especially so given a recent setback suffered by the then popular Ethereum network which led to losses. The Distributed Autonomous Organization (DAO) is a leaderless organization on the Ethereum blockchain (Buntinx, 2016; Tillier, 2016). There is no one in charge of this organization and this DAO became the world's most successful crowdfunding project (Buntinx, 2016; Tillier, 2016). However, the Ethereum network recently lost funds after a code was exploited due to issues with the smart contract technology (Buntinx, 2016; Tillier, 2016). An attacker was able to syphon Ether funds and transfer them to a different address, a dummy version of the DAO, to which only the attacker had the key to (Buntinx, 2016; Tillier, 2016). This attack called Recursive calling vulnerability enabled the hacker to syphon funds repeatedly via a single transaction (Buntinx, 2016).

However, the moderating revelation is that the vulnerability was not from the Ethereum network, but rather from the code used by the DAO developers which reinforces the notion that software-based solutions are not perfect (Buntinx, 2016). The stolen Ether sat in a child DAO where it was locked for some weeks (Buntinx, 2016). Thus, apart from the hacking feat, cashing out the stolen amount was another issue (Buntinx, 2016).

The initial perception was that Bitcoin's slower development was a drawback (Tillier, 2016). Originally, Ethereum advocates touted the ease in making applications for it, but this recent attack has changed that perception (Tillier, 2016). According to Tillier (2016), though Bitcoin does not provide the development capabilities of Ethereum, it has rather shown better reliability (Tillier, 2016). Indeed, this latter statement is better seen in context, since Bitcoin technology has also not been without problems as well.

A couple of years ago (in 2014) an unknown entity exploited a glitch in Bitcoin's core protocol to disrupt the whole Bitcoin ecosystem (Wile, 2014). This problem started at one place and expanded to other major exchanges including BitStamp (the largest in the world) (Wile, 2014). Withdrawals were suspended for a period, but someone thereafter used the same glitch again to syphon Bitcoins belonging to the successor to an illegal online marketplace, The Silk Road (Wile, 2014).

Also recently (in the summer of 2016), hackers stole about $72 million worth of Bitcoins from Bitfinex, a Hong Kong-based crypto-currency exchange (Baldwin, 2016). This hacking is so far the second-biggest security breach of a digital currency exchange (Baldwin, 2016).

Indeed, most of the earlier cited discussions in this paper touted the potential immutable nature of the blockchain technology, but this is better seen from a relative context. Technology moves through phases of development and some of these drawbacks are necessary to improve the systems over time. And just like the use of the internet, over time the potential significant benefits of the technology can outweigh the potential negative outcomes that are likely to accompany the use of such technologies.

Internet hacking and advances in internet security are typical examples of the juxtaposition of the negative and positive trends in the use of influential technologies. Though internet hacking is detestable, most organizations now acknowledge the key role of hacking in certain scenarios. For example, according to Perlroth (2016) most companies in Silicon Valley reward hackers who turn over bugs (i.e., flaws that make products vulnerable for intrusion). Even certain companies such as Apple which was previously against such a policy recently announced that it will pay hackers up to $200,000 to turn over critical and other vulnerabilities in certain Apple products (Perlroth, 2016). These examples are nothing, but significant admission of the role of hacking in technology development. They however pose very deep ethical problems. For example, it may also be argued that a wrong is a wrong and this should not be condoned in any way. In summary, this paragraph provides examples of potential negative versus positive outcomes in the pervasive use of influential technologies (such as the internet) and this can be a useful mirror for reflection in the potential pervasive use of the blockchain technology as well.

IMPLICATIONS FOR PRACTICE

The blockchain technology has certain potential merits and benefits for use in the construction industry to support both project delivery and asset management. Yet, there are a number of significant challenges that need attention to enhance the use of this technology and also reduce risks. The following section provides certain recommendations that can be adopted/adapted to deal with these issues.

First, there is the need for increased awareness in the potential uses, benefits and challenges in the application of the blockchain technology in the construction industry. This awareness can enhance debate and research into the many potential uses of this technology in the construction industry.

Second, national and international collaboration to deal with specific segments of the use of this technology will be useful. This can lead to the provision of guidelines, standards and tools for the eventual adoption of the technology in the construction industry.

Third, there is the need for industry and government collaboration on the use of the technology to improve construction value delivery at both the national and international levels. Governments are often the major buyers of construction services (Barima, 2008b), and such initiatives can provide symbiotic benefits to both parties. Governments can give broad policy frameworks and also support industrial research in this area.

Fourth, a template of the many 'blockchain uses' linked to specific phases of the asset lifecycle as well as specific tactical/strategic goals in the construction industry can be developed to guide progress in this area. This can give a structured framework to guide exploration in this domain. The close multi-disciplinary collaboration among various stakeholders in the construction industry can also be beneficial.

These said, there is always a lag between the emergence of new technologies and their wide acceptance. Also, technologies move through maturity phases. Various allied tools in the use of blockchains may emerge, but vigilance and prudent evolutionary technological adoption/use of the technology in these initial stages will be a useful step. Indeed, some of the touted tools may not even survive over time, and additionally new tools/ideas can also emerge over time to change the industry in a big way.

RECOMMENDATIONS FOR FURTHER RESEARCH

There are a number of opportunities for further research in the leverage of the blockchain technology in the construction industry. They include the following. First, more studies are needed to explore the uses, benefits and challenges in the use of construction specific tools in this domain. This can help accumulate data to inform both practice and research.

Second, as technologies mature in this domain, studies may be needed to distill the business value of these tools in specific areas of construction project delivery and asset management. Since value is a complex construct (Barima, 2010), such studies can help to disentangle the core perceptions of business value in the use of the blockchain technology in the indicated areas.

Third, case studies to document the experience of initial adopters of specific blockchain technologies can both inform practice, support the building of best practices and also inspire learning to improve extant methods and tools.

Fourth, international and national multidisciplinary collaboration by various players in the industry can give synergistic results in the development of standards, best practices, policies, guidelines and knowledge-base to support practice.

Finally, studies on strategic benefits management in the use of the blockchain technology are needed to support technology adoption. Also, on issues dealing with trust in the adoption/use of the technology in the construction industry as it matures over time.

CONCLUSION

This study has illuminated the blockchain technology concept, the potential uses of the technology, as well as the challenges and areas for further studies. Though, the blockchain technology is currently being touted as a potential revolutionary technology with the promise to change extant business operations (including those in the construction industry), beyond this hype there are a number of initial significant challenges that need critical reflection.

The current state of the technology is nascent, and accumulation of significant data/evidence to support the hype is relevant to deal with potential threats that can likely affect the use of such novel technologies in construction business activities. The potential applications of the technology to enhance construction value delivery can be significant, but more studies and an evolutionary approach to technology adoption will be a useful step as the potential uses of this technology move through the initial exploratory phase.

REFERENCES

Allison, I. (2016). Ethereum's Vitalik Buterin asks if Google, Apple and Facebook are threatened by blockchains. *International Business Times*, April 25, 2016. Retrieved July 17, 2016 from http:// www.ibtimes.co.uk/ ethereums-vitalik-buterin-asks-if-google-apple-facebook-are-threatened-by-blockchains-1556543.

Baldwin, C. (2016). Bitfinex says expects 'socialized loss' for $72 million bitcoin hack. Retrieved August 7, 2016 from http://www.reuters.com/ article/us-bitfinex-hacked-hongkong-idUSKCN10G0CZ.

Barima, O. K. B. (2008a). Virtual concept use in the construction industry. In Becker, A. (Ed.), *Electronic commerce: concepts, methodologies, tools, and applications*, (chapter 65, pp. 799-806). Hershey, PA: IGI Global.

Barima, O. K. B. (2008b). E-government and the construction industry. In Anttiroiko, A. (Ed.), *Electronic government: Concepts, methodologies, tools, and applications*, (chapter 243, pp. 3283-3291). Hershey, PA: IGI Global.

Barima, O. K. B. (2010). Examination of the best, analogous, competing terms to describe value in construction projects. *International Journal of Project Management*, *28*(3), 195–200.

BBC, (2016). How block-chain tech could change the way we do business. Retrieved June 20, 2016 from http://www.bbc.com/news/business-35370304.

Buntinx, J. P. (2016). The DAO Loses Ethereum funds through smart contract vulnerability. Retrieved July10, 2016 from http://fintechist.com/dao-loses-ethereum-funds-smart-contract-vulnerability/.

Campbell, R. (2016). University pays $20,000 in Bitcoins to hackers. Retrieved July 7, 2016 from https://www.cryptocoinsnews.com/university-pays-20000-in-bitcoins-to-hackers/.

Castillo, M. (2016). Microsoft launches Blockchain fabric to help enterprises form consortia. Retrieved June 27, 2016 from http:// www.coindesk.com/microsoft-unveils-project-bletchley-streamline-consortium-construction/.

CBC, (2016). University of Calgary paid $20K in ransomware attack. No evidence cyberattackers released personal or university data to public. CBC News, June 7, 2016. Retrieved July 7, 2016 from http://www.cbc.ca/news/canada/calgary/university-calgary-ransomware-cyberattack-1.3620979.

Cellan-Jones, R. (2016). Blockchain and benefits - a dangerous mix? Retrieved July 16, 2016 from http://www.bbc.com/news/technology-36785872+&cd= 2&hl=en&ct=clnk&gl=ca.

Government of Canada, (2016). Construction (NAICS 23): Labour productivity. Retrieved July 31, 2016 from https://strategis.ic.gc.ca/app/scr/sbms/sbb/cis/labourProductivity.html? code=23&lang=eng.

Higgins, S. (2016). US Government awards $600k in grants for blockchain projects. Retrieved July 16, 2016 from http://www.coindesk.com/us-government-grants-blockchain-projects/.

Leising, M. (2016). Central Bankers told they should be sprinting toward blockchain. Retrieved July 16, 2016 from http://www.bloomberg.com/

news/articles/2016-06-06/central-bankers-told-they-should-be-sprinting-toward-blockchain+&cd=2&hl=en&ct=clnk&gl=ca.

Peck, M. E. (2016). Ethereum developer explores the dark side of Bitcoin-inspired technology. *IEEE Spectrum. Institute of Electrical and Electronics Engineers.*, Retrieved June 27, 2016 from http://spectrum.ieee.org/computing/networks/ethereum-developer-explores-the-dark-side-of-bitcoininspired-technology.

Peck, M. E. (2015). The future of the web looks a lot like Bitcoin. *IEEE Spectrum. Institute of Electrical and Electronics Engineers.* Retrieved June 27, 2016 from http://spectrum.ieee.org/ computing/networks/the-future-of-the-web-looks-a-lot-like-bitcoin.

Perlroth, N. (2016). Apple will pay a 'bug bounty' to hackers who report flaws. Retrieved August 7, 2016 from http://www.nytimes.com/2016/08/05/technology/apple-will-pay-a-bug-bounty-to-hackers-who-report-flaws.html?_r=0.

Postscapes, (2016). Tracking the internet of things. Retrieved June 25, 2016 from http://postscapes.com/blockchains-and-the-internet-of-things.

Robles, D. (2015). Smart contract application for the construction industry. *The Ingenesist Project.*, Retrieved July 17, 2016 from http:// www.ingenesist.com/smart-contract-application-for-the-construction-industry/.

Roszak, M. (2016). How bitcoin and blockchain are poised to disrupt CRE. Retrieved June 27, 2016 from http://www.naiop.org/en/Research/About-the-Foundation/June-2016-Newsletter/Bitcoin-and-Blockchain.aspx.

Tillier, M. (2016). Ethereum's woes haven't hurt Bitcoin: Wonder why? Retrieved July 10, 2016 from https://www.cryptocoinsnews.com/ethereums-woes-havent-hurt-bitcoin-wonder/.

Timetric, (2016). Outlook for global construction industry is getting brighter. Retrieved June 9, 2016 from https://timetric.com/info/media-center/press-releases/2016/04/07/outlook-global-construction-industry-getting-brighter/.

Treasury, (2015). Blockchain: explaining the future. Retrieved June 27, 2016 from http://treasurytoday.com/ 2015/09/ blockchain-explaining-the-future-ttti.

Wile, R. (2014). This week's Bitcoin hack was a major setback for hopes of mainstream adoption. Retrieved July 17, 2016 from http:// www.businessinsider.com/bitcoin-hack-setback-2014-2.

Zoe, T. (2016). Does Bitcoin still matter? Retrieved 16 July, 2016 from www.bbc.co.uk/news/technology-36197703+&cd=4&hl=en&ct=clnk&gl=ca.

In: Construction Projects
Editor: Kimberly Hall
ISBN: 978-1-53610-742-5

Chapter 4

'BIG DATA' AND CONSTRUCTION VALUE DELIVERY

Oliver Barima*[*], *PhD
George Brown College, Toronto, Canada

ABSTRACT

The production and use of massive data and information are key intrinsic characteristics of construction value delivery. Therefore, the use of 'big data' techniques/tools can potentially reveal better insights for construction value delivery than the traditional methods applied in dealing with construction data and information.

Yet, the use of 'big data' techniques/tools is a recent concept, and insufficient studies exist in the literature to guide research and practice in the use of such technologies in value delivery in the construction industry. This study fills part of the gap in extant knowledge.

The study extends knowledge by exploring the benefits, challenges and the potential uses of 'big data' techniques and visualization tools in construction value delivery. Additionally, the study provides recommendations for practice and directions for further research. These contributions can help harness the benefits and deal with the threats in the

[*] Corresponding Author Address: 1240 Eglinton Ave West, RPO 37066, Miss.; ON, Canada. Email: obarima7@gmail.com.

use of this novel technological concept (i.e., 'big data') in construction value delivery.

Keywords: big data, construction industry, value delivery, big data analytics, big data, visualization tools, Hadoop

Introduction

Organizational assets exist in many modes. These can be either tangible (e.g., buildings) or intangible (e.g., data) in nature. Unlike the past, where tangible assets were emphasized by businesses, recent trends in business value delivery have shown the tremendous importance of the use of intangible assets in organizational value delivery.

Indeed, one key differentiating factor in value delivery in recent times concerns the demonstration by companies in the beneficial use of intangible assets such as vital insightful information (distilled from data), knowledge and innovation to undertake essential strategic actions. In recent times, technology-based companies such as Facebook, Apple, Google, and Microsoft are a few demonstrated examples of the power behind the proper leverage of such intangible assets in the delivery of corporate growth, influence, market capitalization and real global impact.

Undeniably, contextual differences exist across different industries in any economy. These differences may show in several ways. They include issues such as the competitive environment in which businesses compete; the nature of products and processes ; the maturity level in the use of specific relevant tools (core skills) at specific times; extent of government regulation; general industry practices; level of innovation and technology concentration, etc.

Yet, the proper use of data to distill insightful information, knowledge and innovation, along with the leverage of relevant allied technologies and tangible assets are communally critical across industries. The skillful display of such competencies can essentially play a key role in business value delivery as shown by the successful technology-based companies stated above.

This paper will focus on the construction industry and specifically on one of the identified critical variables (i.e., data). The nature of business processes and products delivered in the construction industry often leads to the generation and use of massive amounts of data from product initiation to asset disposal. This key resource (massive data) is often under-utilized in the construction industry. Recent 'big data' techniques/tools can be used to enhance the proper utilization of these massive construction data to distill insightful knowledge for value delivery in the construction industry.

'Big data' is a burgeoning concept. It concerns the technological use of massive data, often rich in variety, to distill insights for decision making at speeds which are difficult to attain via the use of traditional computing technologies/tools (Keahey, 2013; Manyinka et al., 2011; SAS, 2016; Zikopoulos et al., 2015). Given the novelty of this concept, little attention has been allotted to study the use of pertinent big data techniques/tools to provide potentially better insights in construction value delivery in the literature. This study fills part of the gap in the scanty knowledge on the topic. The paper extends extant knowledge by discussing the potential opportunities, benefits and challenges in the use of 'big data' to improve value delivery in the construction industry. The study also highlights key areas where big data techniques/tools can be used to improve value delivery in the construction industry and moreover it provides areas for further research.

The layout of the paper is as follows. First, the paper describes the nature of the construction industry and the rationale for the use of big data technology in the industry. Second, the study discusses 'big data' as a concept, and also it highlights certain big data techniques/tools applied across industries which have specific relevance in the construction industry. Third, the paper reexamines the construction industry from the lens of 'big data applications' to distill lessons and potential opportunities for use in the construction industry. Finally, the paper provides practical implications, recommendations for further research and conclusions for the study.

BACKGROUND

The Nature of the Construction Industry and the Rationale for the Use of 'Big Data'

The construction industry plays a significant role in global value delivery. The industry usually creates products that are used to create value in other sectors of national economies across the world. Though the construction industry can be described as primarily/ultimately a tangible asset-focused industry, the appropriate use of massive data and information is central to the right delivery of these tangible assets in a safe, environmentally friendly manner at the right price, time and quality in the industry. The proper storage, use of data and information in digital value delivery can provide a useful media for key value delivery improvement in the industry.

Construction data and information are often generated in several ways. One way for understanding this process is the use of the timeline (e.g., the lifecycle of a typical constructed product via a construction delivery method). Typically, constructed products often begin with a vision/idea/concept and progress through the phases of development, implementation, phase-out, operations/maintenance phases till asset disposal.

Each of these identified phases in the construction value delivery process provides an opportunity for the generation, collection, use of massive data and information which can form a good basis for further exploration via the use of big data technologies. Yet, unfortunately most of these construction industry generated data/information exist to a significant extent in hardcopies and they are sometimes discarded for economic and practical reasons such as the need for huge spaces to store such massive data (e.g., Barima, 2008).

Typical projects may generate many types of artifacts (e.g., drawings and pieces of papers (texts), photos, videos, etc.) in a typical construction builder's outfit. For example, a typical picture of the number and types of documents (e.g., drawings and pieces of papers) that may be generated in different types of projects can be as follows:

1. Small projects — 8000.
2. Large projects — 35,000 to 50,000.
3. Certain very large projects — about 100,000 (Barima, 2008).

Additionally, massive information can be generated during the operations and maintenance phases of an asset till disposal. The durable nature of the products delivered by the construction industry makes these latter aspects of the construction industry value delivery processes very important in terms of data gathering.

Indeed, the pervasive use of digital tools to generate, store and use applicable construction data and information can save a lot of physical space and also provide significant economic (e.g., storage), technical, and legal benefits (e.g., Barima, 2008). Additionally, these massive data can form useful basis for improvement in the industry. Opportunities exist now for the industry to distill useful knowledge and insights to inform current and future project delivery, operational/product maintenance and asset disposal decisions at a relative ease and also speed via the application of the relevant technology. This explains why the industry must pay specific attention to recent technological advances in areas such as 'big data.' For example, visualization, analytical techniques and other pertinent tools used in this domain have the potential to enhance better benefits for the industry players. The next section discusses the 'big data' concept from a historical and contemporary perspective and also the discussion focuses on certain tools (techniques and visualization) and some promising technologies that may be of interest to the construction industry.

Big Data Insights and Value Orientation

Big data concerns massive data (by scale) (Keahey, 2013; Manyinka et al., 2011; SAS, 2016; Zikopoulos et al., 2015). The data is often rich in terms of variety, and can be examined with recent applicable big data technologies to reveal latent patterns, correlations and other valuable insights at a coordinated speed and capability which is difficult to do via the use of traditional computing technologies (Keahey, 2013; Manyinka et al., 2011; SAS,2016; Zikopoulos et al., 2015). Data can also be in different forms: structured, semi-structured, unstructured, at rest, or in motion (Zikopoulos et al., 2015).

The concept of using data to gain insights for business decisions is not new (e.g., SAS, 2016). Additionally, certain organizations (e.g., governments, big corporations, etc.) have already been dealing with huge data long before the use of the label 'big data' in the literature (e.g., SAS, 2016). Some of the tools applied in big data analytics are also not new (e.g., statistical correlational analysis tools, etc.). However, what is novel is the speed, efficiency and capability which recent applicable big data technologies can be used to analyze vast data to provide latent actionable insights (Keahey, 2013; Manyinka et al., 2011; SAS, 2016; Zikopoulos et al., 2015).

Indeed, having access to vast data and applicable analytical technologies alone is not enough. Data transformed into relevant analyzed information requires skill to distill insights. Also dexterity is needed to apply any acquired knowledge/insights in a timely strategic or tactical manner to deliver value in new distinct ways.

Hence, a value delivery orientation where interlinked processes in the form of these value adding flow: data-information-knowledge/wisdom-insights-innovation/improvement-timely diffusion within the right strategic/tactical context will be useful to give value to organizations. Additionally, the impact of other eco-systems (by scale and levels) linked to this latter flow should be accounted for to enhance better decision making.

Value is indeed a complex construct, therefore, beyond the current 'big data' hype astute business/technical acumen and value delivery orientation are necessary to deliver or improve value in organizations (Barima, 2010; Zikopoulos et al., 2015).

BIG DATA TECHNIQUES, VISUALIZATION AND PROMISING TECHNOLOGIES

Preamble

Information from 'big data' are not extracted solely via a single technology or tool such as Hadoop (SAS, 2016; Zikopoulos et al., 2015). Rather, a plethora of technologies and techniques (from statistics, computer science especially machine learning, etc.) are applied in tandem to extract the insights from the big data (Keahey, 2013; Manyika et al., 2011; SAS, 2016; Zikopoulos et al., 2015).

Some of the techniques are not entirely new, but the combinations of these techniques are in certain cases intrinsically novel. As an emerging area many new tools will likely evolve over time within their respective technology maturity cycles (see e.g., Press, 2016). This section therefore discusses a sample of some of the typical technologies/tools which can be potentially beneficial to the construction industry. In this section, the analytical techniques used to analyze big data are discussed. Then this is followed by the presentation of pertinent visualization methods used to display big data for insights. Additionally, certain promising big data technological frameworks and allied practices in different phases of maturity are deliberated.

Big Data Techniques

The following provide summarized descriptions of relevant techniques/tools applied in big data analyses (Keahey, 2013; Manyika et al., 2011; Mitchell, 2014; Press, 2016; SAS, 2016):

- Data management is an overarching tool that is relevant to attain well-managed data of high quality and reliability via the maintenance of standards and repeatable processes which can be adapted by an entire organization.

- Data mining: This technique has the capability to distill relevant information from the noise variables in a potentially chaotic data. It combines techniques from machine learning, database management and statistics to distill patterns from large datasets.
- Ensemble learning: Is a type of supervised learning which applies several predictive models (developed from statistics and/or machine learning) to distill improved predictive performance than that which can be delivered by the singular use of each of the constituent models.
- Predictive analytics/modelling is a collection of mathematical models which apply machine learning, data and statistical algorithms/ techniques to predict likely outcomes in the future based on historical data. Currently, relatively cheaper computational power and big data help analysts solve problems that were impossible to do in the past.
- Neural networks: Are models of computation based on biological neural networks used to find patterns in data, and they are especially useful in finding nonlinear patterns. They can be used for pattern recognition and optimization. These techniques can be based on both supervised and unsupervised learning models.
- Optimization techniques: Are a collection of techniques (numerical) applied to redesign complex systems and operational processes (e.g., scheduling) to increase performance based on a set of objective standards (e.g., cost and time). A typical optimization technique example is genetic algorithm.
- Pattern recognition: Is a type of specific algorithm which uses machine learning techniques to identify specific patterns.
- Simulation: Is used for forecasting and scenario planning in complex systems modelling. Monte Carlo simulations, for instance, is a type of algorithm that uses repeated random sampling, to run large numbers of simulations based on different sets of assumptions to give probabilistic outcomes of distributions.
- Signal processing: Is a technique used to analyze discrete and continuous signals (e.g., radio signals, sounds, images) with the capabilities to distinguish/distill signals from noise.
- Spatial analysis: Involves the analyses of topographical, geometric or geographic properties encoded in data sets. Usually these spatial analyses data originate from GIS systems which capture data

including location-based information (e.g., addresses or latitude/longitude coordinates).

- Time series: Concerns the use of forecasting models to predict future trends (e.g., time series) based on known past data of similar/other series. They can be applied to distill seasonal/ cyclical patterns in data.
- Sentiment analysis: Entails application of natural language processing and other analytic techniques to identify and extract subjective information from source text material. The information which can be distilled include new topics and term relations from vast data such as emails, blogs, twitter feeds, surveys and competitive intelligence.

VISUALIZATION TECHNIQUES

The capacity of the human brain assigned to process visual inputs exceeds that for the other human senses (Keahey, 2013). The human visual system has the ability to process megabits (about) of information (Keahey, 2013). This relates to processing about 1 million letters of text per second (Keahey, 2013). Yet, humans face significant challenges in processing enormous amounts of data (numerical and text) (Keahey, 2013; Manyinka et al., 2011).

Traditionally, data visualization in many disciplines have been done with tools such as, line charts, pie charts, bar charts, bubble plots, scatter plots, etc. (Keahey, 2013; Zikopoulos et al., 2015). Whilst these visualization techniques have been useful and fit for their purposes in the past, these tools have significant limitations in terms of visualizing the currently increasing amounts of big data in various industries. Based on Keahey (2013), Table 1 provides a summary of the suggested (rule of thumb) number of items that can be effectively visualized using traditional visualization tools.

Table 1. Items of data which can be shown effectively via the use of traditional visualization tools

Technique	Optimum Number of Items
Pie chart	10
Bar chart	50
Line charts	500
Bubble plots	500
Scatter plots	10,000

Based on the information from Table 1, what will happen if a person has to display data involving say 500,000 items for insights? Indeed, these traditional tools will not be fit for such changing needs in contemporary operational business contexts.

The limitations of traditional tools in visualizing huge data has spawned studies for new visualization techniques to understand such data (e.g., Keahey, 2013; Zikopoulos et al., 2015). 'Visual analytics' has emerged in the literature to support the analyses of big data so as to discern information in big data (Keahey, 2013; Manyinka et al., 2011; Zikopoulos et al., 2015). Analytics help to reduce the size and complexity of big data for practical and effective comprehension (Keahey, 2013; Manyinka et al., 2011; SAS, 2016). Certain advanced techniques in this domain can be used to enhance visualization of up to 1 million items and in special cases about 1 billion items (Keahey, 2013). The integration of visualization and analytic techniques provide a seamless way to obtain the best results from data visualization (Keahey, 2013).

Some of the recent relevant visualization tools that may be applied to big data include tree map, tag cloud, clustergram, map, history flow, spatial flow, and 'river themes.' The following section describes these techniques (Keahey, 2013; Manyinka et al., 2011; Zikopoulos et al., 2015):

- Tree Map: Is an effective and powerful technique to visualize hierarchies. Hierarchy visualization helps to give fuller comprehension of information by showing many hierarchical levels at the same time. In a typical tree map, an outer bigger rectangle represents all the aggregated whole of the hierarchy, and this is in turn subdivided into various aspects of a hierarchy. The size of sub-rectangles and different colors can be used to measure different aspects of a hierarchy. Certain tree visualization techniques can be used to show huge (e.g., hundreds-millions) numbers of entities.
- Tag cloud: Provides quick discernment of the most important concepts in a big body of text data. In tag cloud, words that appear more frequently in a body of text data are displayed in larger words for visualization. Likewise, less frequent words appear as smaller words in the tag cloud.
- Clustergram: Is used in cluster analysis to understand how clustering results change with different numbers of clusters.

- Map: concerns overlaying values from data unto specific geographic maps. They can appear in the form of heat maps to indicate intensity of these representations.
- History flow: Can be used to track the evolution of documents as they are edited by various authors to gain underlying insights on specific issues/trends.
- Spatial information flow: Enhances the spatial flow of information (e.g., Internet Protocol data) across nodes (e.g., specific cities) in the world. The bigger the glow in the spatial flow then the larger the flow.
- 'River themes' visualization: Is a form of technique used in sentiment analysis where the thickness of a band in a displayed data at a point in time relates to the frequency of mentioned words or phrases which indicate the flow and receding ends of sentiments.

Big Data Technologies/Practices in Different Phases of Maturity

The following are some of the promising big data technological frameworks/practices in different phases of maturity (Mitchell, 2014; Press, 2016):

- Search and knowledge discovery: self-service technologies/tools used to extract information/insights from multiple sources (file systems, streams, platforms, applications, databases) of massive structured and unstructured data.
- Data virtualization: technology which provides information from several data sources (e.g., Hadoop, distributed data stores) in real-time and near-real time.
- Stream analytics: software used to filter, combine and analyze high input of data from many distinct live data sources in several data formats.
- Data integration: tools used to orchestrate data across solutions such as Apache Hive, Apache Pig, Apache Spark and Amazon ElasticMapReduce (EMR).
- Data preparation: software that facilitates sourcing, shaping, cleansing diverse and messy data to quicken usefulness of data for analytics.
- Data Quality: products for cleaning and enriching large, high velocity data using parallel operations on distributed data stores and databases.

- Big data Lakes: also referred to as enterprise data hubs are incremental organic models used to build massive databases. Contrary to traditional database theory where a person designs data sets before the entry of data, data lakes do the opposite. Data models are not predesigned before, say, placing them in big Hadoop repositories, and people can dynamically build views into the data over time. The drawback of 'big data lake' is the need for users to be highly skilled.
- Deep learning: This is still evolving. It uses a combination of machine-learning techniques based on neural networking to solve business/other problems. Through deep learning, computers can identify items of interest in massive quantities of unstructured and binary data to deduce relationships without the need for specific models or programming instructions.

Hadoop: is a very important framework in 'Big data,' hence this sub-section is dedicated to discussing some of the elements of Hadoop. Hadoop is an open source software framework with the capacity to store large variable data and also process these data faster via distributed computing model (Keahey, 2013; Manyinka et al., 2011; Zikopoulos et al., 2015). According to Mitchell (2014), Hadoop is the new enterprise data operating system.

Zikopoulos et al. (2015) assert that one of the reasons for the recent increasing interest in Hadoop is that it provides a pragmatic solution to save money and also make money. Through the use of Hadoop, massive data can be stored on lower-tier storage with potential lower software licensing costs to save money (Zikopoulos et al., 2015). Additionally, it can be used to create value by performing analyses of massive variable data that were previously difficult to do (Zikopoulos et al., 2015).

Hadoop has two main aspects: a file system (Hadoop's Distributed File System (HDFS)) and a processing infrastructure (e.g., YARN which also includes the MapReduce framework) (Manyinka et al., 2011; Zikopoulos et al., 2015). They both form part of the Apache Software Foundation (ASF) (Zikopoulos et al., 2015). Additionally, other important Apache projects which work with Hadoop include the following (Apache Software Foundation, 2016; Zikopoulos et al., 2015):

- Hadoop Common: Refers to the common utilities which support other Hadoop components.
- Apache Flume: Is used to gather, combine, and store data in Hadoop.

- Apache HBase: Is a BigTable-style database and it is a real-time read and write database.
- Apache Lucene/Solr: Is a search and indexing tool.
- Apache Hive: Is a tool used to catalogue data stored in Hadoop and also administer queries (SQL-like) on the data.
- Apache Avro: Is a framework for serialization data (transforms data from different formats into binary forms to ensure consistent processing).
- Apache Oozie: Is an engine for scheduling jobs and workflows.
- Apache Pig: Is an advanced data flow language/execution framework.
- Apache Sqoop: Is a tool for data transfer between relational databases and Hadoop.
- Apache Spark: Is a flexible framework for data processing.
- Cassandra: Is a multi-master scalable database that has no single points of failure.
- Mahout: Is a scalable data mining and machine learning library.
- Chukwa: Is a data collection system used for managing large distributed systems.
- Apache Zookeeper: Ensures the functioning of the Hadoop system by monitoring the availability of vital Hadoop services and also managing failovers.
- Ambari: Is a web-based tool used to manage and monitor Apache Hadoop clusters (e.g., including support for Hadoop HDFS, ZooKeeper, Oozie, Pig, etc.). This tool has a dashboard to view the health of the clusters and also diagnose their performance characteristics.

For the initial 7 years of Hadoop's existence MapReduce was the prime distributed processing model for Hadoop (Zikopoulos et al., 2015). However, this tool was very useful for a few specific batch processing (Zikopoulos et al., 2015). In October, 2013, Hadoop 2 was released with a new element called YARN, a general-purpose Hadoop resource manager that permits other applications (processing frameworks) to run on Hadoop clusters in a distributed manner (Zikopoulos et al., 2015). The most popular novel processing framework for Hadoop is the Apache Spark, a more general-purpose framework (than MapReduce) with in-memory processing, that enables programmers to define their own patterns for distributed processing (Zikopoulos et al., 2015).

Hadoop, as a set of tools and a processing framework for analyzing massive data was initially devised to work on physical clusters of machines (Mitchell, 2014). This approach changed and currently several technologies are available to process data in the cloud (Mitchell, 2014).

SQL, MapReduce, in-memory, stream processing, graph analytics and other types of workloads can run adequately on Hadoop and many businesses will likely use Hadoop enterprise data hubs (Mitchell, 2014). Opportunities exist to run various types of queries and data operations against data in Hadoop (Mitchell, 2014). And this will likely make it a low-cost, general-purpose place to put analyzable data (Mitchell, 2014).

Tools that support querying on Hadoop similar to traditional SQL tools enhance ease of use for business users already familiar with SQL (Mitchell, 2014). Additionally, NoSQL (i.e., 'Not Only SQL') databases as alternatives to SQL-based relational databases are increasingly becoming popular tools for specific analytic applications (Mitchell, 2014).

Moreover, there are notable efforts to make Hadoop easy to use for different categories of users without the need for them to possess highly specialized skills in programming needed to do meaningful analyses of data stored in Hadoop. A typical example of such notable efforts is the IBM's BigInsights for Hadoop (BigInsights) which is intended to provide a set of solutions which include the development of integrated set of tools (e.g., 'App store' style interface) for different end users to use different applications (apps) (Zikopoulos et al., 2015). Additionally, it seeks to enhance the integration of existing enterprise IT infrastructure with Hadoop systems (Zikopoulos et al., 2015).

IBM's other contributions also include the following (Zikopoulos et al., 2015):

- *Security:* Enhancing security of the open source Hadoop systems via the BigInsights Web Console structured to act as a gateway to the cluster. The console features enhanced security by supporting Lightweight Directory Access Protocol (LDAP) and Kerberos authentication protocols. BigInsights also supports other authentication solutions such as Linux Pluggable Authentication Modules (PAM). This can, for example, be used to implement biometric authentication and other customized protocols.
- *Flexible File Systems:* This extends the General Parallel File System (GPFS) (which was developed by IBM research in the 1990s) for

Since evidence on the use of big data technologies in specific construction sectors are scanty, this limitation provides good basis for research and observations of actual use of these tools in the industry. This can provide huge opportunities if they are especially linked to past and recent work that have been done in other related areas such as Building Information Modelling (BIM), classifications systems, RFID systems, internet of things, etc.

Techniques such as time series, optimization techniques, simulations, spatial analysis, etc. are already being used in aspects of construction project delivery to enhance value delivery. Some of these techniques (e.g., optimization techniques) have been widely used in construction project management to improve performance in the midst of constraints (e.g., costs& time). Simulations are applied in construction risk management and also in finding optimum design (e.g., architectural, structural, energy) models via the application of BIM. However, opportunities exist for the reuse of these techniques in new ways via exploitation of big data technologies to distill insights from extant massive data at better velocities.

Techniques such as ensemble learning, spatial analysis, neural networks, signal processing and novel data mining tools provide significant promise for improving construction designs, processes and maintenance. Construction is an information intensive industry and the industry has the intrinsic potential access to massive data for big data analyses. Additionally, most construction activities are done directly in the open, implying that environmental variables such as temperature, snow, rainfall, humidity, topography, geotechnical data, etc. are all additional data that can be used for linked analyses of construction inputs, constructed products, construction operations and maintenance. Also, massive data in various forms (e.g., structured, semi-structured, unstructured, at rest and data in motion) can be generated, stored and analyzed daily from many operational non-traditional interconnected sources (e.g., digital parametric BIM models, sensors, RFIDS, smart meters, smart phones and so forth) over the asset lifecycle. Big data analytics can give both linear and non-linear insights to inform decision making in the construction industry over applicable lifecycle phases of construction assets. This can lead to improved products, delivery speed, costs and overall value delivery to applicable players in the industry.

Additionally, visualization tools such as tree maps and clustergrams can support easy visualization of latent information in the analyses of massive data based on specific characteristics of interest. For example, a construction developer of homes can easily use such tools to get insights on customer behavior based on historical data of homes previously sold over the years.

Tools such as tag clouds and history flow can give useful visual insights for legal analyses of discourses that occurred among parties in dispute in construction litigation or arbitration scenarios. Traditionally, access to such insights in disputes on large projects are often daunting, given the huge amounts of potential documentary evidence.

Furthermore, promising technologies/practices (though in different phases of maturity) in areas such as deep learning, Hadoop and allied tools, search and knowledge discovery, data preparation, quality & integration, data virtualization are likely to mature with time to support effective storage and analyses of massive data for insights in the construction industry over time. Construction projects often entail complex activities, sophisticated players, constraints, risks (e.g., intricate environmental conditions) and other variables. The indicated tools can enhance sanity in decision making and also in other areas where machines have the potential to outperform humans.

Also, the potential users of the 'big data' technology in the construction industry can benefit significantly from the open source systems (such as Hadoop) and also additional enhanced work that have been done by organizations such as IBM, Alluxio Open Foundation, and others to enhance the speed, security and also user-friendliness of the open source systems in a cost-effective and transformational manner. These extant efforts can be used as collaborative templates to develop and advance specific tools for use in the construction industry.

Data Management and Policy Issues

In terms of data management, certain construction related organizations have consistently paid attention to data management over the years as part of proper construction business delivery (e.g., Barima, 2008). However, in recent times, information management –based standards, processes, and guidelines for use of digital information and models have gained significant interest in the construction industry (e.g., BRE, 2016; British Standards Institution, 2013; CPIC, 2013). The recent acclaim for the use of BIM in the construction industry has been one key foundation for this latter interest and work (e.g., BRE, 2016; British Standards Institution, 2013; CPIC, 2013). Big data provides another opportunity for the industry to build on these gains to ensure quality, credibility, reliability, relevance and timeliness in the use of massive data/information. This drive in the industry can be supported by big data tools and technologies which deal with data preparation, data quality, and data integration (Press, 2016; Zikopoulos et al., 2015).

To be beneficial international and national collaboration across the industry may be necessary to drive the formation of relevant standards/frameworks for the use of big data technologies to enhance value delivery. This can lead to the building of useful infrastructure for common industry use and also platforms for tailoring to specific private business applications. Indeed, added to these are the unique insights which specific entities/companies/individuals may attain and use to support the overall industry efficiency and aggregate output/supply.

Meaningful policy frameworks (e.g., concerning standards, technology, data, information, legal, commercial issues, etc.) may be relevant at both international and national levels to drive use. These initiatives can enhance common access to beneficial information to all interested parties related to the construction industry at national and international levels of application. Existing works and databases in these areas can form useful bases for the development of such templates where applicable (e.g., BRE, 2016; British Standards Institution, 2013; CPIC, 2013). Indeed, in certain cases extant policy frameworks (e.g., BRE, 2016; British Standards Institution, 2013; CPIC, 2013) in these specific areas (e.g., data/technology) may be just adequate or need minor improvements.

The opportunities for the use of extant templates in the construction industry to support big data applications are vast. For example, ongoing works on classification systems (e.g., Omniclass) exist for different aspects of construction value delivery (e.g., OmniClass Development Committee, 2012). These tools can be used as templates to guide the structured collection of big data for various aspects of the construction industry. It can boost consistency, comparative analyses and aid 'repeatable' use.

Challenges for the Use of 'Big Data' In Construction

The use of the 'big data' concept in the construction industry is not without challenges. First, as a persistent culture, the construction industry often lags behind many of the other industries in technology adoption (e.g., Barima, 2008). Hence, the wider adoption of the concept will entail negotiating a huge initial inertia.

Second, given the novel nature of this domain, there is lack of general awareness of big data applications and benefits in the construction industry. This challenge will require wider dissemination of the potential benefits of big data applications to relevant decision makers in the industry. The success of this process can drive major clients of the construction industry (e.g., national governments) to demand the use/adoption of such technologies by relevant suppliers and other players in the industry.

Third, issues such as initial implementation costs have been noted to be an obstacle to the adoption of big data technologies in other industries (e.g., Accenture, 2014). This budgetary obstacle will be no different in the construction industry, especially given that the industry is fiercely competitive and actors often operate on low margins.

Fourth, computer security has traditionally been a difficult problem for users of interconnected computers and till major breakthroughs occur in this area, this will still continue to be a nagging issue which can lead to potential losses for users of big data technologies. A recent study by Accenture (2014) cited security as one of the key areas of concern in big data application.

Fifth, the significant access of organizations to enormous flood of data in recent times has privacy implications. This can lead to regulatory backlash across nations to limit the collection and use of data, and hence unique insights which can be distilled from them (Tene and Polonetsky, 2013). To deal with this potential threat, there is need to establish a policy balance between individual privacy issues and the constructive applications of data by dealing with the basic concepts of privacy law in areas such as the role of individual control, data minimization and limitation of purpose (Tene and Polonetsky, 2013). Additionally, organizations may have to disclose their decision criteria in relation to inferences drawn from data, since the main source of concern in big data relates to these inferences rather than the data itself (Tene and Polonetsky, 2013).

Finally, for the uses of big data technologies and their benefits to be realized, well prepared human talents are important for fulfilling strategic goals in the use of the 'big data' concept. As the uses of big data technologies grow across the world the availability of skilled personnel in this area is likely to be a huge challenge across all industries (Accenture, 2014) including the construction industry.

FUTURE RESEARCH

The novelty of the 'big data' concept also provides opportunities for future research in the following areas:

1. Studies on the use, risks and benefits of specific big data tools/technology in the industry. Such studies can provide evidence and increased understanding for the strategic and tactical adoption and use of big data tools/technology.
2. Case studies on the use of these technologies by initial adopters detailing the challenges, benefits, and best practices used in such applications. The accumulation of such empirical case studies can significantly boost the proper empirical use of these technologies.
3. Analysis of the cost/benefits or the business value of big data technology/tools in the construction industry. Studies focused in such areas can enhance informed strategic choice and the will of top management in organizations to support the real use of big data technology/tools in the construction industry. Additionally, organizations can learn from such studies to increase the attainable value from the use of big data technology/tools.
4. Research which focus on specific tools and technologies for the use of big data in the construction industry. Also, studies on tools which can specifically enhance construction design (architectural, structural, mechanical, electrical, etc.); additionally, in specific areas in construction processes (e.g., safety, procurement, compliance issues, claims, estimating, dispute resolution, building, erection, equipment usage, improving on materials/ material science) and also operations and maintenance of built facilities and environmental issues.
5. Systemic benefits in the interlinkages between all the different aspects of construction from idea to asset disposal via the use of big data tools. The realignment of research from the fragmented use of big data tools to a systemic inquiry can boost synergistic applications of big data technologies/tools for better value delivery across the construction industry.
6. Policy issues and frameworks (standards, technology, data, legal, commercial, etc.) and also development/documentation of best practices in this domain. As a novel area, this is a very important area

of research to help guide and guard the construction industry from any potential chaos in the use of big data technologies and tools.

7. Construction specific tools to train students and potential users of big data technologies and tools. Such tools can help strengthen the talent pool from which industry can draw on to support the adoption and use of big data technologies to deliver value in the construction industry. Additionally, tailored training of students with construction specific knowledge in areas such as construction materials, science, engineering, architecture, technology, management, facilities management can go a long way to support research and deepen construction practice.

CONCLUSION

This study has extended knowledge on the use of 'big data' as a concept in improving construction value delivery. Just as in other industries, the use of big data technologies/tools has significant promise for the construction industry. The benefits are both generic and specific. The generic business benefits include the ability to gain better insights via the analyses of huge data as well as improved speed in making informed decisions. Others include potential related competitive advantages and cost savings.

Specific benefits in the leverage of the big data concept apply specifically to the improvement of the technical aspects of construction product design, construction processes, operations and maintenance of constructed facilities.

In spite of the benefits, the use of the 'big data' concept also has many challenges. These include the lack of adequate skills in this area; lack of awareness; traditional attitudes in adopting new technologies; security, privacy issues and cost associated with the adoption of big data technologies.

As a novel area sustained communication of the benefits, education/training and further research focused on the application of big data technologies/tools across the construction industry are important to enhance synergistic benefits.

REFERENCES

Accenture (2014). Big success with big data survey. Accenture Report, April 2014.

Alluxio Open Foundation (2016). Open source memory speed virtual distributed storage. Retrieved August 20, 2016 from http://www.alluxio.org.

Apache Software Foundation (2016). What Is Apache Hadoop? Retrieved August 3, 2016 from http://hadoop.apache.org.

Barima, O. K. B. (2008). Virtual concept use in the construction industry. In Becker, A. (Ed.), *Electronic commerce: concepts, methodologies, tools, and* applications (chapter 65, pp. 799-806). Hershey, PA: IGI Global.

Barima, O. K. B. (2010). Examination of the best, analogous, competing terms to describe value in construction projects. *International Journal of Project Management 28*(3), 195–200.

BRE (2016). Level 2 requirements. Building Research Establishment Ltd. Retrieved May 20, 2016 from http://www.bre.co.uk/page.jsp?id=3447.

British Standards Institution (2013). PAS 1192-2: 2013. BSI Standards Limited.

CPIC (2013). Post contract-award Building Information Modelling (BIM) Execution Plan (BEP). Construction Project Information Committee. Retrieved April 28, 2016 from http://www.cpic.org.uk/cpix/cpix-bim-execution-plan/

Keahey, T. A. (2013). Using visualization to understand big data. IBM Corporation.

Li, H.(2016). Introducing Alluxio open source project governance. Retrieved August 14, 2016 from http://www.alluxio.com/2016/06/introducing-alluxio-open-source-project-governance/.

Li, X., Xu, L. and Lv, X.(2016). Qunar performs real-time data analytics up to 300x faster with Alluxio. Retrieved August 14, 2016 from http://www.alluxio.com/2016/07/qunar-performs-real-time-data-analytics-up-to-300x-faster-with-alluxio/#more-661.

Manyika, J., Chui, M., Brown, B., Bughin, J., Dobbs, R., Roxburgh, C., Byers, A. H. (2011). Big data: The next frontier for innovation, competition, and productivity. McKinsey Global Institute.

Mitchell, R. (2014). Big data technologies and practices are moving quickly. Here's what you need to know to stay ahead of the game. Retrieved June

26, 2016 from http://www.computerworld.com/article/2690856/big-data/8-big-trends-in-big-data-analytics.html.

OmniClass Development Committee (2012). OmniClass: A strategy for classifying the built environment. OmniClass Development Committee.

Press, G.(2016). Top 10 hot big data technologies. Forbes, March 14, 2016.

SAS (2016). Data Visualization: What it is and why it matters. Retrieved June 1, 2016 from http://www.sas.com/en_sg/insights/big-data/data-visualization.html.

Tene, O. and Polonetsky, J.(2013). *Big data for all: privacy and user control in the age of analytics. Northwestern Journal of Technology & Intellectual Property 11* (5) 239. Retrieved June 25, 2016 from http:// scholarlycommons.law.northwestern.edu/njtip/vol11/iss5/1.

Timetric (2016). Outlook for global construction industry is getting brighter. Retrieved June 9, 2016 from https://timetric.com/info/media-center/press-releases/2016/04/07/outlook-global-construction-industry-getting-brighter/.

Zikopoulos, P. deRoos, D., Bienko, C., Buglio,R., & Andrews, M.(2015). *Big data beyond the hype: A Guide to conversations for today's data center.* New York: McGraw-Hill Education.

In: Construction Projects
Editor: Kimberly Hall
ISBN: 978-1-53610-742-5

Chapter 5

MITIGATION OF CONSTRUCTION DISPUTES THROUGH RELATIONAL CONTRACTING IN PUBLIC PROJECTS IN HONG KONG

Caroline T. W. Chan, PhD*
Division of Building Science and Technology,
City University of Hong Kong, Hong Kong, P.R.C.

ABSTRACT

To cope with the growing population over the years, the Hong Kong government spends more than HK$70 billion in infrastructural and institutional developments annually. Being the largest client in the construction industry, the government strives to complete the public projects in a timely and cost effective manner. However, the lowest-bid tendering method and the increasingly complex project nature raise a large number of construction disputes in the public projects. In 2009, the government piloted the use of NEC3 contract (a form of relational contract) in a few civil engineering projects, attempting to alleviate the problem of construction disputes by establishing partnership with the contractors. The projects were satisfactorily completed, with time and cost savings to the government. More importantly, the number of disputes in each project was substantially reduced as most of the construction or interfacing problems were resolved once they became apparent.

* Corresponding Author: caroline.chan@cityu.edu.hk.

Although building partnerships at work has been promoted for a long time, distrust between the project parties is deep-rooted. The conventional way of non-contractual partnership brings very little improvement in dispute mitigation. Successes in the NEC3 projects provide valuable information for us to design and implement relational contracts in order to minimise construction disputes. Data were collected from a questionnaire survey and follow-up interviews with 83 project participants who were involved in three projects using NEC3 contracts. The key attributes that promote partnership in a relational contract were identified and the effectiveness of partnership developed through relational contract attributes was cross-examined with Macneil's relational contract norms. Project stakeholders can use the recommendations to promote collaboration and minimise disputes through relational contracting.

1. Introduction

1.1. Public Sector of the Hong Kong Construction Market

Since the handover of Hong Kong to Mainland China in 1997, the population of Hong Kong has increased from 6.5 million to 7.3 million (HKSAR, 2016a). To provide a better living environment for the growing population and to enhance the economic competitiveness of the city, the Hong Kong government has initiated a "Ten Major Infrastructure Projects" plan, which covered the construction of Kai Tak Development (that is, the redevelopment of the old Kai Tak Airport area), the Hong Kong section of the Guangzhou-Shenzhen-Hong Kong Express Rail Link, the Hong Kong-Zhuhai-Macau Bridge, the Rail connection between the Hong Kong and Shenzhen airports, Tuen Mun-Chek Lap Kok Link, Lok Ma Chau Loop, the West Kowloon Cultural District, the North East New Territories New Development Areas, the South Island Line and the Sha Tin to Central Link of the Mass Transit Railway extensions. These ten projects are complemented with a series of sewage and waste treatment plants, municipal and environmental facilities, heritage conversation projects and many others to sustain the long-term development of Hong Kong. These mega projects pushed the public works expenditure to HK$77.1 billion in 2015, increased by 12% in nominal terms when compared with 2014 (HKSAR, 2016b).

1.2. The Hong Kong Government and NEC3

Being the owner of many capital works projects, the Hong Kong government is committed to seek improvements in the project delivery process, viz., to complete the projects within the shortest time frame, the lowest cost and to the highest quality standard. In the Report on Construct for Excellence, the committee strongly advocated "the integration of a partnering approach into the contractual relationship by developing a new form of contract that is based on co-operation, client focus and commitment to best practice for application to local construction project" (Construction Industry Review Committee, 2001). In response to the CIRC's recommendation, the government piloted partnering in a few projects and most project teams agreed that partnering was worthwhile for implementation.

Over the past decade, the New Engineering Contract (NEC3), which is advocated to enhance partnership effectively in construction contracts, has attracted much attention in Europe. The Hong Kong government embarked its first trial of NEC3 contract in 2009 in a sewage improvement project. Since then, a few more NEC projects have been piloted. The experience gained in the pilot runs on NEC3 provided a strong foundation for the government to use NEC3 in all public works contracts as far as possible starting from 2015. Up to July 2016, there are eight NEC3 public projects completed and the results are positive. All of them were completed on time and within budget, with one of them finished earlier and at a final cost lower than the target cost.

2. Overview of NEC3 Contract

2.1. Structure of NEC3 Contract

The first edition of the NEC, called the New Engineering Contract, was published in 1993 by the Institution of Civil Engineers (ICE), U.K. Two years later the second edition, NEC Engineering and Construction Contract (ECC), was published. The third edition, NEC3, was launched in 2005 and further expanded into a suite of 39 documents in 2013. As listed in Table 1, the NEC3 suite covers a complete set of principal contract, short contracts, sub-contracts, professional services contract, supply contracts and adjudicator's contract. These contracts are interlocked using the same phraseology so that project owners (or employers) can use them to procure contractors and consultants for design, fabrication and construction.

Table 1. Documents within the NEC3 suite

Contracts/subcontracts	Guidance notes and flow charts	Other supporting documents
NEC3 Engineering and Construction Contract (ECC)*	Engineering and Construction Contract Guidance Notes	Procurement and Contract Strategies
ECC Option A: Priced contract with activity schedule	Engineering and Construction Contract Flow Charts	How to write the ECC Works Information
ECC Option B: Priced contract with bill of quantities	Engineering and Construction Short Contract Guidance Notes and Flow Charts	How to use the ECC communication forms
ECC Option C: Target contract with activity schedule	Professional Services Contract Guidance Notes and Flow Charts	How to write the PSC Scope
ECC Option D: Target contract with bill of quantities	Professional Services Short Contract Guidance Notes and Flow Charts	How to use the PSC communication forms
ECC Option E: Cost reimbursable contract	Term Service Contract Guidance Notes	How to write the TSC Service Information
ECC Option F: Management contract	Term Service Contract Flow Charts	How to use the TSC communication forms
NEC3 Engineering and Construction Subcontract (ECS)	Term Service Short Contract Guidance Notes and Flow Charts	How to use BIM with NEC3 Contracts
NEC3 Engineering and Construction Short Contract (ECSC)	Supply Contract Guidance Notes	
NEC3 Engineering and Construction Short Subcontract (ECSS)	Supply Contract Flow Charts	
NEC3 Professional Services Contract (PSC)	Supply Short Contract Guidance Notes and Flow Charts	
NEC3 Professional Services Short Contract (PSSC)	Framework Contract Guidance Notes and Flow Charts	
NEC3 Term Service Contract (TSC)	Adjudicator's Contract Guidance Notes and Flow Charts	
NEC3 Term Service Short Contract (TSSC)		
NEC3 Supply Contract (SC)		
NEC3 Supply Short Contract (SSC)		
NEC3 Framework Contract (FC)		
NEC3 Adjudicator's Contract (AC)		

*Note: Structure of the NEC3 Engineering and Construction Contract (ECC) is illustrated further in Figure 1.

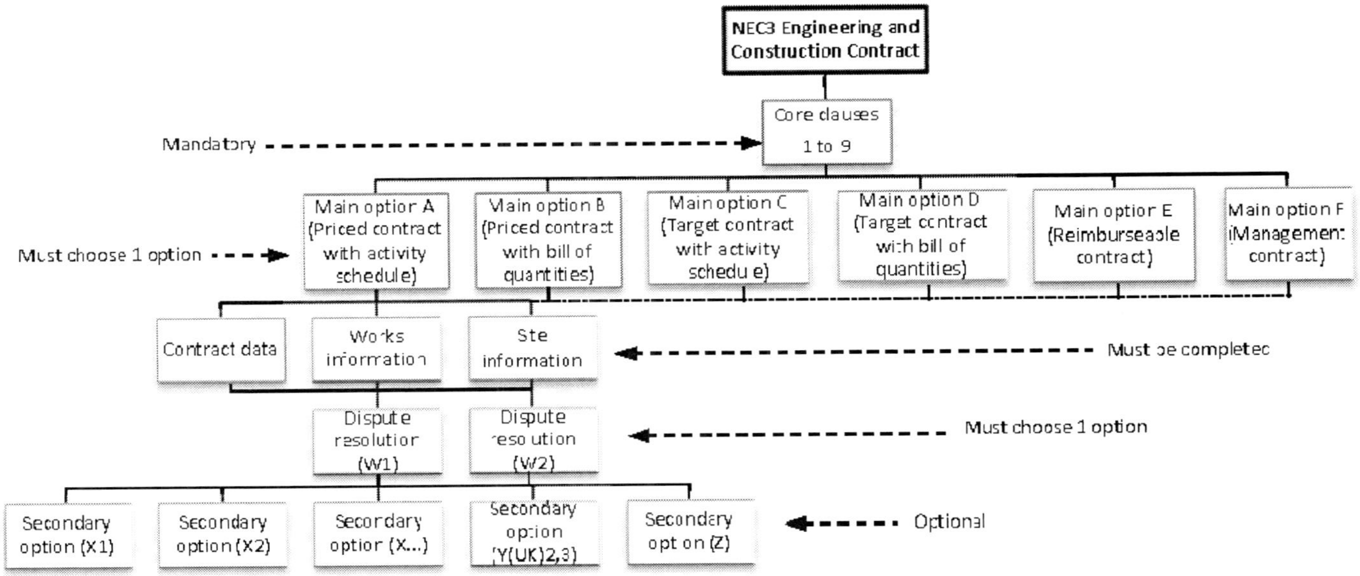

Figure 1. Structure of the NEC3 Engineering and Construction Contract (ECC).

Considering the procurement of the main contractor for a construction project, NEC3 Engineering and Construction Contract (ECC) is the most popular choice. Therefore, ECC is chosen to explain further. Figure 1 describes the various components in the ECC. The base contract consists of nine core clauses[1]. On top of these core clauses, the employer can choose one of the six payment options (option A to option F) including:

- Option A: lump sum contract based on activity schedule defined and priced by contractor
- Option B: remeasurement contract, priced with bill of quantities
- Option C: target cost contract, target costs set via activity schedule
- Option D: target cost contract, target costs set via bill of quantities
- Option E: cost reimbursable contract
- Option F: management contract

Contract data, works information and site information are project-specific details that must be clearly described in the contract. The employer has a further choice of secondary options[2] to adjust the risk allocation between the contract parties.

2.2. Key features of ECC (NEC3)

Many scholars describe NEC3 as a form of collaborative contract (Baker, 2006; Manu et al., 2015) or relational contract (El-adaway, 2013; Kan and Le, 2014). The ethos of collaboration is core to the NEC3 suite. One of the key objectives for the design of NEC contracts was to stimulate good management and to reduce the risks inherent in construction by foresighted, co-operative management of the interactions between the parties (Institution of Civil Engineers, 2013). The salient features of the ECC are summarised as below.

[1] The nine core clauses are: 1. General; 2. The Contractor's main responsibilities; 3. Time; 4. Testing and Defects; 5. Payment; 6. Compensation events; 7. Title; 8. Risks and insurance; and 9. Termination.

[2] Examples of secondary options: price adjustment option (X1), parent company guarantee option (X4), perfomance bond option (X13) and bespoke clauses option (Z).

(A1) Use of plain, simple English

As distinct from traditional contracts, the ECC is written in plain English, in the present tense and with minimal use of legal terminology. Concise sentences are used throughout the contract. Bullet points and lists are used effectively to break up the text.

(A2) Flexible structure

The ECC is not sector specific. The contract can be applied to any form of building, civil engineering and mechanical engineering projects. The ECC is designed in a modular fashion, to avert the need for different payment methods and many bespoke changes. The document contains a choice of main options that offer different forms of pricing, risk allocation and payment mechanisms. The extent of the contractor's design obligations, which can be whole or specific part of the works, is described in the Works Information. In addition, the ECC is coupled with dispute resolution options and a series of secondary options to suit different project requirements. The employer can also insert additional conditions of contract in the Clause Z to cater for specific needs.

(A3) Codify collaboration

As stipulated in the first core clause (Clause 10.1) of the ECC, all project participants should act as stated in the contract and in a spirit of mutual trust and co-operation. This clause obliges all parties to do openly and honestly everything which the contract states they should (Institution of Civil Engineers, 2013).

(A4) Pain/gain sharing

The pain/gain sharing mechanism is available in options C and D (target cost contracts). The principle that underlies the target cost contract is that a target cost is agreed between the employer and the contractor. Throughout the course of the work, the contractor is reimbursed for the costs he expends plus an agreed fee. Upon completion of the project, the final actual expenditure (the Defined Cost) spent by the contractor, together with the fee, is compared to the final target cost. If the final actual cost plus fee is less than the target cost, the "saving" is shared between the employer and the contractor on a pre-agreed percentage basis. This is referred to as gain share. On the contrary, if the actual cost plus fee is higher than the target cost, the "loss" is shared between the two parties according to a pre-agreed ratio, commonly known as pain share. The target cost options provide financial incentives to the contractors, encouraging them to perform their duties effectively. This explains why option C (target cost) is the most popular form in the NEC3 (Mason, 2016).

(A5) Competency of Project Manager

The Project Manager is the core party acting as the employer's agent, to ensure that all the employer's objectives are met. It is the Project Manager who issues instructions; accepts subcontractors; accepts programmes; reviews and accepts design and method statements; assesses compensation events; maintains risk register; and assesses amounts due for payment. Although in practice the role of Project Manager is often undertaken by a multi-disciplinary team, the competency of the Project Manager cannot be understated. The obligation to work in a spirit of mutual trust and co-operation applies to the Project Manager as well. The Project Manager has to act independently, impartially and honestly when accepting or rejecting designs (Clause 21.2), assessing compensation events (Clause 62.3) and accepting or rejecting programmes (Clause 31.3). Also, the strict timescales laid down in the contract require the Project Manager to perform proactively and respond on time.

(A6) Works information

Under the ECC, Works Information is the specification and description of works and constraints that the contractor must conform to, such as access restrictions and materials storage requirements. The prime purpose of the Works Information is to describe what is to be done on the site by the contractor. Works Information is divided into two parts: the employer's Works Information that includes drawings, specifications and constraints; and the contractor's Works Information that describes the particulars of the contractor's design. The main feature of the Works Information is that if there is any employer's design change or constraint not included in the Works Information, these amendments can be made later by a Project Manager's instruction, which will then constitute a compensation event (Clause 60.1).

(A7) Strict timescales

Strict time provisions are set for contract communications, programme revisions as well as assessment of payments and compensation events. However, the Project Manager and the contractor can agree to extend some timeframes if they have difficulty to abide by the timescales.

(A8) to (A10) Risk management

ECC encourages proactive risk identification and joint problem solving through *risk register* (A8), *early warning notifications* (A9) and *risk reduction meetings* (A10). Project parties are obligated to give notification to the other party if they foresee any matter that could increase the project cost, delay the completion/key date or impair the work performance (Clause 16.1). The raised early warning notifications will be followed up in risk reduction meetings. These meetings aim to identify proposals to deal with and mitigate the notified risks. Clause 16.3 makes it clear that the parties in these meetings must co-operate to find solutions that will bring advantage to all affected parties.

(A11) Detailed and regularly updated programme

ECC lays down clear provisions on the submission of the first and revised programmes. Timing and the information required on each programme are stated clearly in Clause 31.

(A12) Compensation events assessment

Under the ECC, the procedures for claiming additional time and payment have been consolidated into the compensation event procedures. The contractor is reimbursed according to the actual Defined Cost for work already done plus the fee; and the forecast Defined Cost for work not yet done plus the fee. Therefore, the whole assessment is based on the effect of the compensation event on the contractor's cost. Any over- or underpricing in the tender will not affect the assessment of the compensation event.

(A13) Interim payments assessment

Unlike traditional contract practice, every interim certificate is final, not an on account payment. The purpose of this assessment approach is to maintain a rolling final account for both parties. Besides, when assessing payment under options C and D, the Project Manager should forecast the amount of Defined Cost that the contractor will incur before the next assessment date. This gives the contractor a better cash position when compared with traditional contract arrangement.

(A14) Mandatory adjudication

Dispute resolution in all NEC contracts operates in a two-tier approach, first by adjudication, then subsequently through appeals to the tribunal (arbitration or litigation as selected by the contracting parties). All adjudication expenses, including the adjudicator's fees, are shared equally between the two parties irrespective of the outcome.

(A15) Procurement of subcontractors

Under Clause 26, the Project Manager has the authority to accept (or reject) a subcontractor. The contractor must obtain the Project Manager's acceptance of the proposed subcontractor and the subcontract terms.

3. MACNEIL'S RELATIONAL CONTRACT THEORY

3.1. Relational Contract Theory

Relational contract is a form of agreement that is sustained by mutual trust and benefits through collaborative relationship between the contracting parties. The nature of relational contract has been investigated by scholars from the economics (Baker et al., 2002; Pilling *et al.*, 1994), marketing (Dwyer et al., 1987; Ivens and Blois, 2004) and legal disciplines (Eisenberg, 1999; Gordon, 1985). Most of these studies are theoretical in nature and regarded as relational contract theory. Among them, the norm-based approach developed by Macneil is widely considered as the fundamental theory for the analysis of relational contracts (Campbell et al., 2013; Morgan, 2013).

Macneil describes a contract as "no more or less than the relations among parties to the process of project exchange into the future" (Macneil, 1980), and therefore a contract exists whenever there is exchange. Essentially, Macneil believes that all contracts involve co-operative social behaviour - "….the kind of social behaviour involved is co-operative social behaviour, behaviour characterised by a willingness and ability to work with others …. Contract involves people affirmatively working together." (Macneil, 1968).

3.2. Common Contract Norms

A norm is "a principle of right action binding upon the members of a group and serving to guide, control, or regulate proper and acceptable behaviour" (Macneil, 1980). Macneil developed a set of ten common contract norms that can be found in different forms of contract (Macneil, 1980). These common norms underpin all contracts by fostering a cooperative attitude to varying degrees, producing a spectrum of "contractual relations" from relational to discrete.

The ten common contract norms are:

(N1) Role integrity: refers to a contracting party's behaviour to maintain a long-term relationship by trusting the other party to behave properly and fulfil each other's obligations.

(N2) Reciprocity: relates to the reasonable anticipation of a contracting party that his/her counterparty would give back something that is commensurate with his/her own contribution in the contractual relationship.

(N3) Flexibility: relates to contracting parties' ability to reconstruct the content of their relationship and their readiness to compromise on ambiguities.

(N4) Contractual solidarity: is the norm that holds the exchange together, causing the contracting parties to adopt actions that maintain a stable and continuous relationship.

(N5) The "linking norms": include restitution; reliance and expectation interest. Restitution comes into play when one contracting party gains an unfair benefit; reliance gives effect to non-legally binding promises and expectation interest refers to what has been promised. Macneil described them as linking norms because they often link the other contract norms.

(N6) Creation and restraint of power (the "power" norm): is the existence of power creation of a party on another, as well as the ability to restrain such power for the benefit of the other party.

(N7) Propriety of means: relates to the parties' possession of adequate means to perform their obligations.

(N8) Harmonisation of the social matrix: refers to the influence of social principles such as justice, liberty, trust and equality on the contracting parties.

(N9) Implementation of planning: refers to the adoption of plans and measures to set out how things are done and how relationships in the contract are structured.

(N10) Effectuation of consent: relates to the exercise of a choice, which thereby sacrifices other opportunities. In case of a contract, the exercise of a choice often means that the power to restrain a party's future choices is vested on the other party.

Macneil further explained that the significance of each contract norm varies in different forms of exchanges. For instance, discrete contracts are characterised by the intensification of two contract norms, namely *implementation of planning* (N9) and *effectuation of consent* (N10). Macneil described them as "enhancing discreteness and presentation." The two norms are now commonly regarded as the discrete norms (Beale et al., 2007; Campbell et al., 2013; Mitchell, 2013).

In addition to the ten common contract norms, Macneil (1983) suggested that other relational norms often exist within a relationship. These supplementary relational norms are *mutuality* and *open communication.*

(N11) Mutuality: refers to the "fair" distribution of the gains and losses between the contracting parties.

(N12) Open communication: relates to sharing of open communication and sensitive information.

4. Knowledge Gap and Present Initiative

Although NEC3 and relational contract implementation attracts much discussion in the industry over the past years, the literature to date has focused primarily on the identification of critical factors leading to successful collaboration (Black et al., 2000; Chen and Chen, 2007; Doloi, 2015; Jefferies et al., 2014; Liu et al., 2014; Zou et al., 2014) and the benefits of using NEC contract (Construction Industry Council, 2015; Fullalove, 2014; Hassan et al., 2013; Wilsoncroft, 2009; Wright and Fergusson, 2009). Past research on the prerequisites for relational contracts to be successful has been limited. The eight Hong Kong public projects that have been recently completed through ECC provide a good foundation for the study of relational contracting. Complementing Macneil's relational contract theory with an empirical approach, the efficacy of relational ties developed through ECC contracts is investigated. In addition, the key relational attributes in the ECC and their effectiveness to reduce construction disputes are explored, which provide useful findings for wider consideration.

5. Method

To investigate the effectiveness of using ECC as a relational contract to mitigate construction disputes, both quantitative and qualitative data were collected. The approach consisted of a questionnaire survey as the primary source of quantitative data to gain insight into such practices, and follow-up interviews to gather qualitative views on the subject.

5.1. Research Framework for the Study

5.1.1. Relational Scores

The framework for evaluating the strength of relational norms in the ECC is shown in Figure 2. The model used only Macneil's relational norms (including eight common relational norms N1 to N8 and two supplementary relational norms N11 and N12) as the relational criteria, each of which linked to fifteen ECC attributes. These ECC attributes were identified through literature review followed by pre-survey interviews with six construction professionals. The critical ECC features that were emphasised in the NEC3 literature or mentioned by four or more interviewees were selected as the 'ECC attributes' for the model.

The strength of the relational norms in each ECC attribute was then measured. Each relational criterion (under each ECC attribute) was rated a score ranging from 0 to 5, representing "no impact" to "very strong" relationship between the parties. A relational score that reflected the strength of relationship would be calculated for each ECC attribute, which was the mean of scores earned for each relational criterion. This enables an objective comparison of the relational qualities across the ECC attributes.

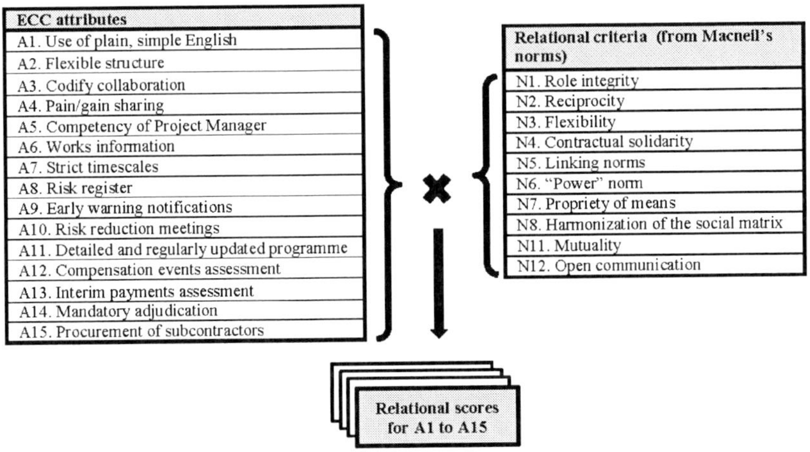

Figure 2. Framework of indicators for evaluating relationship.

5.1.2. Effectiveness of Dispute Mitigation

To analyse the effectiveness of EEC to mitigate disputes, the same list of ECC attributes (Table 2) was used. The perceived effectiveness of dispute mitigation for each ECC attribute was measured to evaluate the overall performance of the ECC contract.

Table 2. ECC attributes for dispute mitigation study

	ECC attributes
A1.	Use of plain, simple English
A2.	Flexible structure
A3.	Codify collaboration
A4.	Pain/gain sharing
A5.	Competency of Project Manager
A6.	Works information
A7.	Strict timescales
A8.	Risk register
A9.	Early warning notifications
A10.	Risk reduction meetings
A11.	Detailed and regularly updated programme
A12.	Compensation events assessment
A13.	Interim payments assessment
A14.	Mandatory adjudication
A15.	Procurement of subcontractors

5.1.3. Correlation between Relational Scores and Effectiveness of Dispute Mitigation

Since the relational scores and the effectiveness of dispute mitigation were measured against the same set of ECC attributes, correlation between the two can be tested for each ECC attribute.

5.2. Questionnaire Design

The aim of the questionnaire was to collect opinions on the strength of relationship developed through the ECC attributes and the effectiveness of

ECC to mitigate disputes. To achieve the objective, the questionnaire was designed with three sections. The first section collected demographic data from the respondents. The second section evaluated the strength of relational norms present in the selected ECC attributes by requesting the respondents to assign a score of 0 (no impact) to 5 (very strong) to each ECC-relational criterion. As mentioned before, the ECC attributes were generated from literature review and pre-survey interviews whereas the relational criteria were compiled from Macneil's relational norms. The final section asked the respondents to indicate the effectiveness of each ECC attribute to mitigate construction disputes, on an effectiveness scale of 0 (no impact) to 5 (highly effective). The questionnaire was pilot tested before implementation to enhance validity.

5.3. Sample Design

Due to the limited number of knowledgeable practitioners on NEC contracts, purposive sampling was adopted. Government departments and construction companies that are involved in NEC3 projects were sourced from the internet, within which project managers and senior engineers were targeted. Eleven practitioners were identified as being accessible and were contacted, out of which eight availed themselves. These eight practitioners came from three NEC3 public projects. To increase the number of responses, seventy-five additional candidates were identified by snowball sampling. The candidates recruited by snowball were project staff of the three NEC3 projects, covering representatives from the employers (government departments), main contractors and consultants. A total of eighty-three completely filled questionnaires were obtained.

After analysis of the survey data, ten respondents were approached, out of which eight agreed to participate in the follow-up interview. These follow-up interviews were arranged to obtain a better understanding of the opinions given by the respondents in the questionnaire survey. By asking probing questions during the interviews, the respondents were required to provide in-depth answers whereby peculiar aspects of their responses could be delved into.

6. Survey Results

6.1. Respondent Profiles

The eighty-three respondents were classified into three categories: employers (government departments) (17%), main contractors (59%) and consultants (24%). The average work experience of the respondents is 6.5 years, with about 18% of them having more than 10 years of experience. Table 3 summarises the details of the ECC projects engaged by the respondents.

Table 3. ECC projects engaged by the respondents

	Project 1	Project 2	Project 3
Contract type	ECC Option C	ECC Option C	ECC Option C
Nature of work	Drainage improvement	Highway improvement	Drainage works
Contract sum (HK$)	76 million	138 million	678 million
Client	Government	Government	Government
Project manager	Government	Government	Government
Supervisor	External consultant A	External consultant B	Government
Quantity Surveying consultant	External consultant A	External consultant B	External consultant C
Experience of partnering between the employer and the contractor	None	None	Yes
Completion date	2012	2013	2017
Early completion	6 months	Nil	Not yet available
Cost saving	5%	Nil	10% (estimated)

6.2. Results

Data analysis was done by the Microsoft Office Professional Plus 2010 – Excel. Descriptive statistics dealt mainly with the ranking of the variables based on mean values and standard deviations. Analysis of variance (ANOVA) was used to test the null hypothesis that mean values of the dependent variables are the same for all respondent groups. Correlation analysis between the relational scores and the effectiveness of dispute mitigation was conducted to identify if there is a relationship.

6.2.1. Reliability Analysis

Reliability analysis was applied to test the internal consistency of the survey questions. Cronbach's alpha coefficients are 0.820 ($p < 0.001$) for the questions that measure the strength of relational norms in ECC and 0.867 ($p < 0.001$) for those that measure the effectiveness of ECC to mitigate construction disputes. Both coefficients are greater than the 0.70 threshold and therefore, the questionnaire should suffice as a reliable instrument for the study (Reynold and Santos, 1999).

6.2.2. Mean Traits

Table 4 and Table 5 show the descriptive statistics of the survey results, with the ECC attributes arranged in descending order of their mean scores. As the probability (p) values for the one-way ANOVA are greater than 0.05, there is not enough evidence to reject the null hypothesis that the means for all groups are the same at the significance level $\alpha = 0.05$.

Table 4. Descriptive statistics for the relational scores of ECC attributes

Rank	Ref.	ECC attributes	Relational score	Standard deviation	ANOVA	
					F-ratio	p-value
1	A4	Pain/gain sharing	4.146	0.260	2.211	0.116
2	A3	Codify collaboration	4.135	0.334	3.102	0.050
3	A9	Early warning notifications	4.089	0.283	0.625	0.538
4	A10	Risk reduction meetings	4.069	0.285	0.057	0.945
5	A11	Detailed and regularly updated programme	4.058	0.310	0.376	0.688
6	A13	Interim payments assessment	4.027	0.292	0.130	0.878
7	A7	Strict timescales	4.014	0.291	0.510	0.602
8	A8	Risk register	4.010	0.323	0.037	0.964
9	A12	Compensation events assessment	4.003	0.285	0.098	0.907
10	A6	Works information	3.982	0.455	0.276	0.759
11	A2	Flexible structure	3.977	0.309	0.074	0.929
12	A1	Use of plain, simple English	3.966	0.369	0.308	0.735
13	A5	Competency of Project Manager	3.919	0.414	1.370	0.260
14	A14	Mandatory adjudication	3.851	0.407	0.046	0.955
15	A15	Procurement of subcontractors	3.304	0.344	1.570	0.214

For analysis purpose, the scores of "3" were considered as "medium" on a 5-point scale. Therefore, any mean scores below "3" were regarded as "low" whereas above "3" were treated as "high." Table 4 shows that all ECC attributes yielded above 3.0 in the relational scores, indicating that the ECC is highly effective to foster strong relational ties between the project parties. The five ECC attributes that were ranked highest were *pain/gain sharing* (A4), *codify collaboration* (A3), *early warning notifications* (A9), *risk reduction meetings* (A10) and *detailed and regularly updated programme* (A11). The *pain/gain sharing* (A4) attribute ranked the first place with the highest mean score (4.146) and the lowest standard deviation (0.260). Nine out of fifteen ECC attributes scored higher than 4.0. Except for *codify collaboration* (A3), all the top-ranking attributes were related to risk and financial management. The least relational attribute was *procurement of subcontractors* (A15), with the lowest score of 3.304.

Table 5. Descriptive statistics for the effectiveness of ECC attributes to mitigate disputes

Rank	Ref	ECC attributes	Mean score	Standard deviation	ANOVA	
					F-ratio	p-value
1	A9	Early warning notifications	4.365	0.266	0.794	0.455
2	A10	Risk reduction meetings	4.293	0.262	0.069	0.934
3	A11	Detailed and regularly updated programme	4.293	0.330	0.977	0.381
4	A5	Competency of Project Manager	4.281	0.457	1.490	0.231
5	A4	Pain/gain sharing	4.245	0.280	0.077	0.926
6	A8	Risk register	4.136	0.203	0.173	0.842
7	A7	Strict timescales	4.127	0.350	0.084	0.920
8	A12	Compensation events assessment	4.088	0.205	0.810	0.449
9	A3	Codify collaboration	4.019	0.452	0.019	0.981
10	A1	Use of plain, simple English	3.771	0.379	0.141	0.868
11	A13	Interim payments assessment	3.335	0.385	1.794	0.173
12	A6	Works information	3.251	0.243	0.341	0.712
13	A14	Mandatory adjudication	3.227	0.220	0.099	0.906
14	A2	Flexible structure	2.954	0.207	2.593	0.081
15	A15	Procurement of subcontractors	2.711	0.199	0.546	0.581

As Table 5 illustrates, the majority of the respondents thought that ECC was effective to mitigate construction disputes. For the fifteen ECC attributes offered to respondents, the mean scores ranged from 4.365 to 2.711. The five attributes that were the most effective to mitigate construction disputes were *early warning notifications* (A9), *risk reduction meetings* (A10), *detailed and regularly updated programme* (A11), *competency of Project Manager* (A5) and *pain/gain sharing* (A4). Their top scores (ranged from 4.365 to 4.245) were much higher than the top five relational scores (from 4.146 to 4.058).

Four of the highest-ranked ECC attributes, including *pain/gain sharing* (A4), *early warning notifications* (A9), *risk reduction meetings* (A10) and *detailed and regularly updated programme* (A11), were the same in Tables 4 and 5. Interestingly, *competency of Project Manager* (A5) was considered poorer or weaker in terms of its ability to promote trust and collaboration (at rank 13 in Table 4) but very effective to mitigate disputes (at rank 4 in Table 5). Two attributes, namely *flexible structure* (A2) and *procurement of subcontractors* (A15), were considered "low" in their effectiveness to mitigate disputes. *Procurement of subcontractors* (A15) was the least effective in promoting relationship and reducing disputes. This indicates that the prior approval of subcontractors by the Project Manager as stipulated in the ECC contract cannot reduce potential risks effectively.

6.2.3. Frequency Analysis

To clarify the efficacy of ECC features in developing and sustaining the relationship among the contracting parties, the relational criteria that evidenced a high strength (i.e., those given a rating higher than "3" by more than 85% of the respondents) are denoted by an "X" in Table 6. The cut-off line was set at 85% of the respondents so that a reasonable number of relational criteria could be filtered for further examination.

An analysis of the relational effectiveness across ECC attributes indicated that the majority of the ECC attributes significantly supported at least one relational criterion. As presented in Table 6, *pain/gain sharing* (A4) was the most effective contract feature to develop relationship, especially in terms of *contractual solidarity* (N4), *linking norms* (N5), *propriety of means* (N7), *harmonisation of the social matrix* (N8) and *mutuality* (N11) among the project team members. The second most effective attribute, *codify collaboration* (A3), was found to be strong in promoting *role integrity* (N1), *propriety of means* (N7) and *harmonisation of the social matrix* (N8). These agree with the high relational scores for the *pain/gain sharing* (A4) and the *codify collaboration* (A3) attributes. The rest of the ECC attributes were more effective in developing at most two relational criteria.

From Table 6, five ECC attributes strongly promoted *open communication* (N12). Half of the relational criteria were significantly supported by two to three ECC attributes. These relational criteria or norms include *flexibility* (N3), *contractual solidarity* (N4), *propriety of means* (N7), *harmonisation of the social matrix* (N8) and *mutuality* (N11). The rest of the four relational criteria were supported largely by a single ECC attribute.

Table 6. Effectiveness of ECC attributes with respect to relational criteria

Ref	Relational criteria / ECC attributes	Relational score	N1 Role integrity	N2 Reciprocity	N3 Flexibility	N4 Contractual solidarity	N5 Linking norms	N6 "Power" norm	N7 Propriety of means	N8 Harmonisation of the social matrix	N11 Mutuality	N12 Open communication
N1	Pain/gain sharing	4.346				X	X		X	X	X	
N2	Codify collaboration	4.135	X						X	X		
N3	Early warning notifications	4.089										X
N4	Risk reduction meetings	4.069			X							X
N5	Detailed and regularly updated programme	4.058										X
N6	Interim payments assessment	4.027		X							X	
N7	Strict timescales	4.014										
N8	Risk register	4.010										X
N9	Compensation events assessment	4.003									X	X
N10	Works information	3.982			X							
N11	Flexible structure	3.977										
N12	Use of plain, simple English	3.966				X						
N13	Competency of Project Manager	3.919										
N14	Mandatory adjudication	3.851						X				
N15	Procurement of sub-contractors	3.304										
Frequency			1	1	2	2	1	1	2	2	3	5

Note: Relational criteria with more than 85% of the respondents rated 4 or 5 are indicated by an 'X.'

6.2.4. Correlation Analysis

The relationships between the relational score variable and the dispute mitigation variable were investigated using Pearson's correlation analysis. The correlation coefficients in Table 7 indicate that a positive association exists between relational strength and dispute mitigation effectiveness. Such results confirm that the strength of relational norms increases with an increase in dispute mitigation effectiveness. This finding is in accordance with the expectation, since more collaborative teams can normally solve problems more efficiently, which should lead to less disputes between the parties.

Table 7. Pearson's correlation between relational norms and dispute mitigation

Rank	Ref	ECC attributes	Correlation (*r*)
1	A4	Pain/gain sharing	0.603
2	A1	Use of plain, simple English	0.597
3	A15	Procurement of subcontractors	0.593
4	A10	Risk reduction meetings	0.580
5	A8	Risk register	0.577
6	A6	Works information	0.576
7	A13	Interim payments assessment	0.573
8	A12	Compensation events assessment	0.547
9	A7	Strict timescales	0.541
10	A14	Mandatory adjudication	0.536
11	A11	Detailed and regularly updated programme	0533
12	A9	Early warning notifications	0.526
13	A2	Flexible structure	0.510
14	A3	Codify collaboration	0.320
15	A5	Competency of Project Manager	0.354

Notes: All correlations are significant at the 0.01 level (2-tailed).

The highest correlation value was for *pain/gain sharing* (A4) ($r = 0.603$, $p < 0.01$), showing the strongest correlation between relational score and dispute mitigation effectiveness. According to Cleff (2013), if *r* lies between 0.5 to 0.8, there is moderate correlation. Results show that association between relational strength and dispute mitigation effectiveness for most ECC attributes, including *pain/gain sharing* (A4), was moderate. Low, positive but significant correlation can be seen for *codify collaboration* (A3) and *competency of Project Manager* (A5). This is likely due to the fact that both features are strong in one variable but relatively weaker in the other variable. In case of *codify collaboration* (A3), although Clause 10.1[1] of the ECC enhances the commitment in collaboration, it has less direct impact on dispute mitigation. For the *competency of Project Manager* (A5), it is understood that the ECC places a strong demand on making prompt responses and decisions by the Project Manager in order to mitigate disputes. However, in practice, many Project Managers are reluctant to make decisions swiftly especially when assessing compensation events. This may lead to distrust between the contractor and the Project Manager, which then impedes the relationship development between the two parties.

7. DISCUSSION

Many government projects nowadays are mega-sized and complex, financed by a huge amount of public funds. Fostering collaborative project teams to complete the projects effectively is the principal objective of the public works departments in the government. This study affirms the existence of strong relational norms in the NEC3 (ECC) contract, which is likely to promote trust and team spirit among the team members. The survey results also evidence that the ECC mitigates disputes effectively. Several issues emerge from the survey findings are elaborated below.

7.1. ECC (NEC3) Effectively Promotes Collaboration

Despite the recent debates on whether good faith clauses (such as Clause 10.1 in ECC) are enforceable in court (Court, 2012; Willan, 2014), many empirical studies have proved the success of NEC3 to foster collaboration (McLernon and McClements, 2015; Mickovski et al., 2013; Wright and Fergusson, 2009). The relational scores in Table 4 imply that the ECC can foster co-operation effectively. This finding is validated by the "partnering evaluation" data collected from the post-survey interviews. According to the interviewees from Project 3, effectiveness of collaboration among the team members was monitored by the stakeholders on a bi-monthly basis.

[1] Clause 10.1 in ECC states that all project parties "shall act as stated in this contract and in a spirit of mutual trust and co-operation."

"Partnering evaluation surveys" were conducted with selected managerial staff and front-line staff on site to assess the perceived level of trust commitment and co-operation on several aspects (see Table 8 and Table 9 respectively).

Table 8. Aspects of measurement for the trust commitment scores

	Trust commitment aspects
1	Punctual/quick response
2	Openness/communicate
3	Honour
4	Considerate
5	Understanding and fair

Table 9. Aspects of measurement for the co-operation scores

	Co-operation aspects
1	Good planning
2	Joint ownership
3	Responsibility
4	Sharing of information
5	Caring for each other
6	Empathy and passion

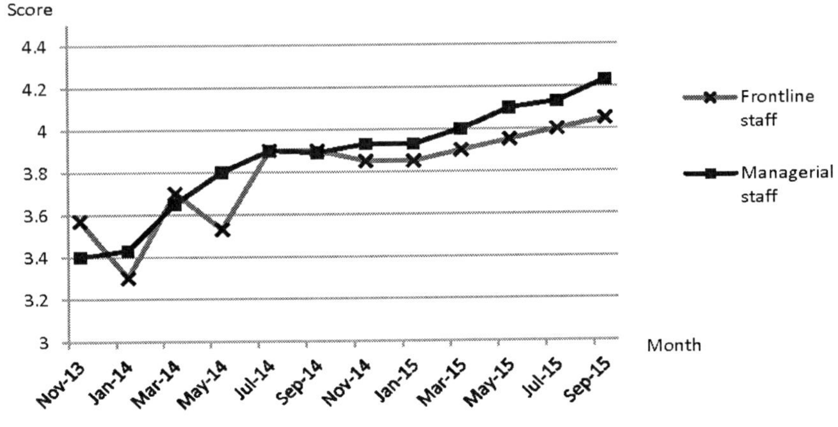

Figure 3. Trust commitment scores from front-line staff and managerial staff.

Evaluation results were then distributed and reviewed during the stakeholders meetings. Figure 3 and Figure 4 show the trust commitment scores and the co-operation scores respectively. A climbing trend can be seen in both figures, indicating that collaboration within the team had been improved since project start. These findings align with previous research that empirically validated the benefits of NEC3.

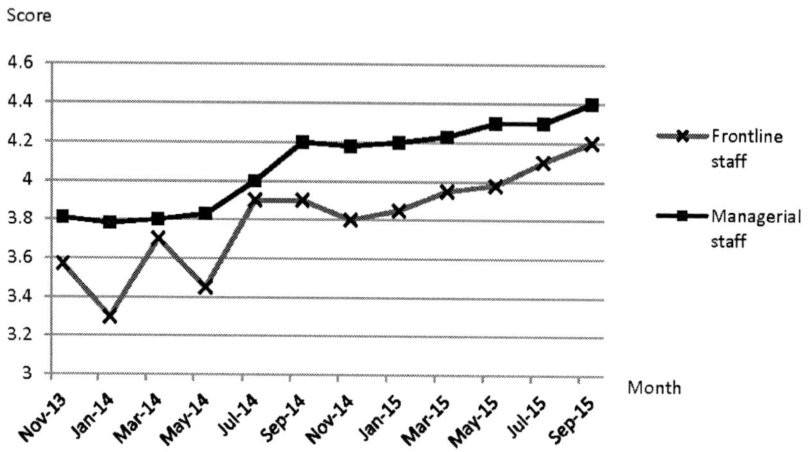

Figure 4. Co-operation scores from front-line staff and managerial staff.

7.2. *Pain/Gain Sharing* as the Keystone of the Team

Tables 4 and 6 show that the most important feature of the ECC is the *pain/gain sharing* (A4) mechanism. Follow-up interviews reveal that the *pain/gain sharing* mechanism in the ECC target cost contract provides direct incentives to the parties to collaborate in order to share the savings. This shared objective promotes a win-win attitude among the project team members (Johnson et al., 2011; Lord et al., 2010). Macneil (1980) emphasised that a relational contract must provide mutual benefit, to bear some burdens and collect some benefits, to both parties (*mutuality* (N11)). The mutual benefit will stimulate interdependence between the two parties (*reciprocity* (N2)) and commitment to preserve the relationship (*contractual solidarity* (N4)). The three projects involved by the respondents were NEC3 target cost projects (Table 3) with the employer's pain capped at 110% - 120% but no limit for the gain share. Interestingly, survey and interview results indicate that all parties including the contractors considered the mutual benefit and risk allocation as equitable, albeit the pain and gain was not equally shared between them. This can be explained by Macneil (1980) that the division of mutual benefit needs not to be equal in order to develop the *mutuality* norm (N11). The key is whether both parties perceive the mutual benefit as "enough" rather than "equal."

A few interviewees also commented that the *pain/gain sharing* (A4) mechanism together with the prescribed payment assessment method in the ECC guaranteed the amount payable to the contractor. This "expectation of return" contributes well to the *linking norms* (N5) (Macneil, 1980). *Pain/gain sharing* (A4) mechanism is also recognised to promote innovative thinking and to reward outstanding performance (Koodie, 2009; Pishdad-Bozorgi and Karasulu, 2013; Scheepbouwer and Humphries, 2011). This eventually improves the contractor's possession of adequate means to perform the job (*propriety of means* (N7)). Under the *pain/gain sharing* (A4) system, both the contractor and the employer are more willing to consider alternative methods or designs as they can share the gain if an alternative method works. For instance, interviewees from Project 3 highlighted that a number of innovative proposals were raised by the contractor during the construction stage. The most representative one was a foundation redesign proposal, using a raft foundation with sub-surface drainage system in lieu of 533 pre-bored H piles. The revised design resulted in HK$70 million cost saving and 5-month time saving.

All interviewees unanimously agreed that sustainability, fairness, equitable sharing of return, collaboration, peace and harmony were important social norms, which could be realised in target cost projects. Not surprisingly, *harmonisation with social matrix* (N8) was rated high by the respondents for the *pain/gain sharing* (A4) and the *codify collaboration* (A3) attributes (Table 6). If the contract can embody social norms and values and achieve a harmonisation, the relationship and trust can be intensified, enabling the parties to absorb conflict and to solve problems by their own community (O'Toole, 2007).

7.3. *Codify Collaboration* Encourages Workplace Friendships

Survey results indicate that *codify collaboration* (A3) was the second most important feature in the ECC (Tables 4 and 6). All interviewees concurred that Clause 10.1 literally brought everyone in the project to work in a spirit of mutual trust and co-operation. It was achieved in two ways. First, front-line staff considered this as a social norm as well as a mandatory obligation (which explains why *harmonisation of social matrix* (N8) for *codify collaboration* (A3) scored high in Table 6). Second, senior management of the team implemented different means to foster the "spirit of mutual trust and co-operation." To enhance the team spirit, a team logo was designed and used for all marketing communications. Joint site office was adopted for the contractor and the employer's representatives. Various kinds of workshops were arranged on site to equip staff with the necessary partnering skills and NEC knowledge. In addition, partnering meetings and stakeholders meetings were held at regular intervals to review the "partnering evaluation" results and to seek improvement in co-operation. These training and meetings encourage the team to think out of the box, thereby increasing the means to perform their obligations (*propriety of means* (N7)). On the other hand, co-worker friendships are proven to enhance trust, respect and co-operation among the team members (Morrison, 2004; Nielsen et al., 2000; Yang et al., 2012). Companies can contribute to develop such social relationships or friendships by arranging workgroups or dinners, for instance (Fine, 1986; Riordan and Griffeth, 1995). In Project 3, led by the senior management of the contractor, community services including school visits and fun fairs were co-arranged with the employer to promote the functions and duties of the government department. Furthermore, social activities such as dragon boat racing, birthday parties and annual dinners were arranged for the whole project team. The workplace relationships were effectively extended to the social relationships among the team members. These arrangements sustain the long-term relationship and mutual trust between the parties with respect to the *role integrity* norm (N1) (Macneil, 1980). Ongoing workplace friendships also encourage mutual understanding and *open communication* (N12) over time.

7.4. Risk Management through *Open Communication*

Table 6 shows that *open communication* (N12) is the strongest relational criteria in the ECC. Interviewees explained that in order to achieve open communication, joint site office was used for all the contracting parties. All interviewees agreed that the colocation of the team enhanced collaboration between the designers and the contractor.

Open and effective communication (N12) is deliberated as highly essential in relational development. Macneil (1978) considered formal and informal communication as "extensive and important" in relational exchange. Specifically, Cheng et al. (2000), Hartmann (2006), Low (1999) and Wong and Cheung (2004) identified the importance of open communication in achieving successful relationships among the project parties. However, a relationship is unlikely to flourish unless there is bilateral communication (Dwyer et al., 1987). As shown in Table 6, all the risk management attributes were considered as strongly relational with respect to the *open communication* norm (N12). This is because ECC embeds a comprehensive communication system in the risk management procedures, starting from the *risk register* (A8), *early warning notifications* (A9), *risk reduction meetings* (A10) and finally to the *compensation events assessment* (A12). *Detailed and regularly updated programmes* (A11) intertwine with the above processes to ensure sufficient amount of details are included in the programmes for review and action. The entire risk management system emphasises on the use of face-to-face meetings (*risk reduction meetings* (A10)) and two-way standard forms (for *risk register* (A8) and *early warning notifications* (A9)) to exchange project information between the Project Manager and the contractor. Such open sharing of information and ideas helps to resolve the risks and align the project objectives among the project parties (Chan *et al.*, 2004). Eventually, mutual trust is strengthened, which helps to forge good relationships.

7.5. ECC (NEC3) Effectively Mitigates Construction Disputes

From Table 5, the perceived effectiveness of ECC to mitigate construction disputes was notably high. This finding is further supported by the information collected from the post-survey interviews. Details of the *early warning notices* and the *risk reduction meetings* from the three projects were summarised in Table 9. In the completed projects, more than 90% of the potential risks could be resolved effectively through the *early warning notification* system (A9) and the *risk reduction meetings* (A10).

Table 9. Summary of the number early warnings and risk reduction meetings

Projects	Number of early warning notices issued	Number of risk reduction meetings held	Number of early warning notices resolved	Status
1	15	11	15 (100%)	Completed
2	28	18	26 (93%)	Completed
3	41	49	36 (88%)	On-going

These results replicate those of past studies (Hassan et al., 2013; Li et al., 2016; Williams et al., 2013; Wilsoncroft, 2009), confirm that ECC introduces a comprehensive risk management system to mitigate disputes. In general, the *use of plain, simple English* (A1) throughout the ECC achieves clarity and simplicity for readers. The risk management system (A8 to A10) in the ECC encourages the parties to proactively and collaboratively identify potential problems, and to seek ways to mitigate them. Both the contractor and the Project Manager are obliged to notify the risks (using *early warning notifications* (A9)) as soon as the risks become aware of a matter that could affect time, cost or quality of the project. The early warning notice (A9) under ECC gives a right to the parties to order a *risk reduction meeting* (A10), to identify and agree the mitigation measures. Subsequently, the Project Manager considers whether there are any compensation events as a result of his instructions or notifications of compensation events. In order to ensure the above procedures are stringently carried out by the parties, *strict timescales* (A7) are set in the contract. Further, the *detailed and regularly updated programmes* (A11) reflect the actual progress and the impacts of compensation events on a timely basis. All these ECC attributes (that scored above 3.5 in Table 5) promote early settlement of disputes effectively.

7.6. Applying Relational Contract in Practice

Relational contracts provide the means for sustaining mutual trust and partnership between the contracting parties. The success lies predominantly on the equitable share of benefits and risks. *Pain/gain sharing* (A4) mechanism provides a strong incentive for the team to co-operate. While conventional contracts tend to focus on the blame allocation when having problems, relational contracts attempt to mitigate problems by proactive early warnings (A9) and collaborative problem solving. However, this requires a change of mindset among all the project parties. The employer's representatives, including the designers and the cost administrator, should be empowered to make swift and bold decisions in order to maintain mutual trust and collaboration within the team. Workshops and training related to the technical and the managerial aspects of the relational contract should be provided to all key members of the team.

The effectiveness of relational contracting relies on the proactive attitude of team, which has to be supported by sufficient amount of resources. Employer and contractor should not underestimate the additional resources required to achieve effective project administration. Noting that heavy workload on documentation is required for the open book accounting system, target cost contracts will be more applicable in middle-sized to large-sized projects that employ large contractors.

LIMITATIONS OF THE STUDY

Due to the lack of relational projects completed in Hong Kong, the study was based on the data collected from three public projects only. As such, the empirical findings of this study may not represent the views of the entire construction industry. Care should be taken when projecting the findings to other sectors or regions.

RECOMMENDATIONS FOR FUTURE RESEARCH

Further studies can be done in other regions where relational contracts are more prevalent, so as to generate a larger database for comparison among different regions or cultures. It would also be interesting to explore how the collaborative relationships develop further downstream of the subcontracting tiers.

CONCLUSION

Contractual disputes are expensive, time-consuming, and detrimental to the relationship between the employer and the contractor. By adopting relational contracting, contractual disputes can be minimised. This study examines the use of ECC (NEC3) contract in three public projects to encourage co-operation and to mitigate construction disputes. The empirical results from a survey of 83 NEC project participants in Hong Kong indicate that ECC can effectively promote relational ties and minimise disputes. The *pain/gain sharing* (A4) and the risk management features, including *compensation events assessment* (A12), *early warning notifications* (A9), *risk reduction meetings* (A10) and *detailed and regularly updated programme* (A11) of the ECC were the most effective in promoting co-operation. A similar pattern of results emerged with regard to dispute mitigation. Strength of relationship was positively correlated with effectiveness of dispute mitigation for most of the ECC attributes. By testing Macneil's relational norms in the ECC public projects, *open communication* (N12) and *mutuality* (N11) were the most prominent in relationship development. The findings of this study assist researchers and practitioners in the construction industry to design relational contracts for public projects. Interested parties can focus their attention on developing the most critical relational norms in order to sustain the collaborative relationship among the team members.

REFERENCES

Baker, E., 2006. Partnering Strategies: The Legal Dimension. [WWW Document]. URL http://jp.whitecase.com/files/Publication/b3db0334-8709-4dba-8be6-65e1c81497cf/ Presentation/PublicationAttachment/b9426d17-52d4-4f87-809a-6bcad3557574/ Partnering_Strategies_The_ Legal_Dimension.pdf.

Baker, G., Gibbons, R., Murphy, K. J., 2002. Relational contracts and the theory of the firm. *The Quarterly Journal of Economics,* 117(1), 39–84.

Beale, H. G., Bishop, W. D., Furmston, M. P., 2007. Contract, Cases and Materials, 5th Edition. *Oxford University Press*, Oxford.

Black, C., Akintoye, A., Fitzgerald, E., 2000. An analysis of success factors and benefits of partnering in construction. *International Journal of Project Management,* 18(6), 423–434.

Campbell, D., Mulcahy, L., Wheeler, S. (Eds.), 2013. Changing Concepts of Contract: Essays in Honour of Ian Macneil. *Palgrave Macmillan*, Hampshire.

Chan, A. P. C., Chan, D. W. M., Chiang, Y. H., Tang, B. S., Chan, E. H. W., Ho, K. S. K., 2004. Exploring critical success factors for partnering in construction projects. *Journal of Construction Engineering and Management,* 130(2), 188–198.

Chen, W. T., Chen, T. T., 2007. Critical success factors for construction partnering in Taiwan. *International Journal of Project Management*, 25(5), 475–484.

Cheng, E. W. L., Li, H., Love, P. E. D., 2000. Establishment of critical success factors for construction partnering. *Journal of Management in Engineering,* 16(2), 84–92.

Cleff, T., 2013. Exploratory Data Analysis in Business and Economics: An Introduction Using SPSS, Stata, and Excel. *Springer International Publishing*, Switzerland.

Construction Industry Council, 2015. Frequently Asked Questions on NEC3 Collaborative Contracts.

Construction Industry Review Committee, 2001. Construct for Excellence, Report of the Construction Industry Review Committee. Hong Kong.

Court, J., 2012. Good Faith clauses: do they mean anything? [WWW Document]. URL http://www.kwm.com/en/es/knowledge/insights/good-faith-clauses-do-they-mean-anything-20121106.

Doloi, H., 2015. Key factors of relational partnerships in project management, in: Schwindt, C., Zimmermann, J. (Eds.), Handbook on Project Management and Scheduling, International Handbooks on Information Systems, Volume 2. *Springer International Publishing*, Berlin, pp. 1047–1061.

Dwyer, F. R., Schurr, P. H., Oh, S., 1987. Developing buyer-seller relationships. *Journal of Marketing,* 51(2), 11–27.

Eisenberg, M. A., 1999. Why there is no law of relational contracts. *Northwestern University Law Review,* 94, 805.

El-adaway, I. H., 2013. Promoting the sustainability of relational contracting through addressing third party insurance obstacles. *Journal of Management in Engineering,* 29(3), 216–223.

Fine, G. A., 1986. Friendships in the work place, in: Derlega, V.J., Winstead, B. A. (Eds.), Friendship and Social Interaction, Springer Series in Social Psychology. *Springer*, New York, pp. 185–206.

Fullalove, S., 2014. Hong Kong Government's NEC Journey: An Interview with Wai Chi-Sing. *NEC Users' Group Newsletter,* 68, 2–3.

Gordon, R. W., 1985. Macaulay, Macneil, and the discovery of solidarity and power in contract law. *Wisconsin Law Review,* 1985, 565.

Hartmann, A., 2006. The role of organizational culture in motivating innovative behaviour in construction firms. *Construction Innovation*, 6, 159–172.

Hassan, S., Bashir, L., Bashir, S., 2013. Implementation and benefits of NEC3 form of contracts in Indian construction industry. *International Journal of Management, IT and Engineering,* 3(11), 90–111.

HKSAR, Census and Statistics Department, 2016a. Population. [WWW Document]. URL http://www.censtatd.gov.hk/hkstat/sub/so20.jsp.

HKSAR, Census and Statistics Department, 2016b. Construction output for the fourth quarter and the whole year of 2015. [WWW Document]. URL http://www.censtatd.gov.hk/press_release/pressReleaseDetail.jsp?charsetID=1&pressRID=3818.

Institution of Civil Engineers, U.K., 2013. Guidance Notes - NEC3 Engineering and Construction Contract (ECC), 3rd Edition, Reprinted with amendments 2013. *Thomas Telford*, U.K.

Ivens, B. S., Blois, K. J., 2004. Relational exchange norms in marketing: A critical review of Macneil's contribution. *Marketing Theory,* 4(3), 239–263.

Jefferies, M., Brewer, G. J., Gajendran, T., 2014. Using a case study approach to identify critical success factors for alliance contracting. *Engineering, Construction and Architectural Management,* 21(5), 465–480.

Johnson, T. R., Sitzabee, W. E., Feng, P. P., 2011. Evolution of relational contracting in construction: project delivery methods beyond partnering. *procs*: 16th ICCRTS International Command and Control Research and Technology Symposium. 21-23 June 2011, Quebec, Canada.

Kan, H., Le, Y., 2014. Research on contract governance in the field of project management. *procs*: the 2014 International Conference on Construction and Real Estate Management (ICCREM 2014), American Society of Civil Engineers. 27-28 September 2014, Kunming, China, pp. 733–741.

Koodie, A., 2009. Performance partnering and key performance indicators (KPIs) - A UK perspective. *procs*: WEFTEC 82nd Annual Water Environment Federation Annual Technical Exhibition and Conference, Water Environment Federation. 10-14 October 2009, Orlando, U.S., pp. 6470–6479.

Li, D., Lo, W., Shou, S., 2016. Delivering the Cheung Chau submarine pipeline in Hong Kong using NEC3, *Proceedings of the Institution of Civil Engineers - Civil Engineering*, 1–6.

Liu, H., Wang, M. J., Skibniewski, M. J., He, J. S., Zhang, Z. S., 2014. Identification of critical success factors for construction innovation: From the perspective of strategic cooperation. *Frontiers of Engineering Management,* 1(2), 202–209.

Lord, W., Liu, A., Tuuli, M. M., Zhang, S., 2010. A modern contract: Developments in the UK and China. *Proceedings of the Institution of Civil Engineers: Management, Procurement and Law,* 163, 151–159.

Low, S. P., 1999. The extension of construction partnering for relationship marketing. *Marketing Intelligence and Planning* 17(3), 155–162.

Macneil, I. R., 1983. Values in Contract: Internal and external. *Northwestern University Law Review*, 78, 340–418.

Macneil, I. R., 1980. The New Social Contract: An Inquiry into Modern Contractual Relations. *Yale University Press,* New Haven.

Macneil, I. R., 1978. Contracts: Adjustment of Long-term Economic Relations under Classical, Neoclassical, and Relational Contract Law. *Northwestern University Law Review.* 72(6), 854–902.

Macneil, I. R., 1968. Contracts, instruments for social cooperation, East Africa: Text, cases, materials. *F.B. Rothman*, U.S.

Manu, E., Ankrah, N., Chinyio, E., Proverbs, D., 2015. Trust influencing factors in main contractor and subcontractor relationships during projects. *International Journal of Project Management,* 33(7), 1495–1508.

McLernon, T., McClements, S., 2015. Comparing and evaluating the contractual provisions of two analogous construction contracts for the management of cost. *procs*: COBRA Research Conference 2015. 8 - 10 July 2015, Sydney, Australia.

Mickovski, S. B., Black, J. D., Smith, M. J., 2013. Innovative use of ECC (NEC3) for procurement and management of infrastructure projects with limited funding: Bervie Braes case study, in: Smith, S. D., Ahiaga-Dagbui, D. D. (Eds.). *procs*: 29th Annual ARCOM Conference. 2 - 4 September 2013, Reading, U. K., pp. 799–808.

Mitchell, C., 2013. Contract Law and Contract Practice: Bridging the Gap Between Legal Reasoning and Commercial Expectation. *Hart Publishing*, Oxford.

Morgan, J., 2013. Contract Law Minimalism: A Formalist Restatement of Commercial Contract Law. *Cambridge University Press*, U. K.

Morrison, R., 2004. Informal relationships in the workplace: associations with job satisfaction, organisational commitment and turnover intentions. *New Zealand Journal of Psychology* 33(3), 114–128.

Nielsen, I. K., Jex, S. M., Adams, G. A., 2000. Development and validation of scores on a two-dimensional workplace friendship scale. *Educational and Psychological Measurement,* 60(4), 628–643.

O'Toole, T., 2007. Defining close relationships among relationships. *procs*: 23rd Industrial Marketing and Purchasing (IMP) Group Conference. 30 August - 1 September 2007, Manchester, U. K.

Pilling, B. K., Crosby, L. A., Jackson, D. W., 1994. Relational bonds in industrial exchange: An experimental test of the transaction cost economic framework. *Journal of Business Research*, 30(3), 237–251.

Pishdad-Bozorgi, P., Karasulu, Y., 2013. Advancing target price and target value design process in IPD using BIM and risk-sharing approaches.

procs: 49th ASC Annual International Conference. 10-13 April 2013, San Luis Obispo, California.

Reynold, J., Santos, A., 1999. Cronbach's Alpha: A tool for assessing the reliability of scales. *The Journal of Extension,* 37(2), 1-5.

Riordan, C. M., Griffeth, R. W., 1995. The opportunity for friendship in the workplace: An underexplored construct. *Journal of Business and Psychology,* 10(2), 141–154.

Scheepbouwer, E., Humphries, A., 2011. Transition in adopting project delivery method with early contractor involvement. *Transportation Research Record: Journal of the Transportation Research Board*, 2228, 44–50.

Willan, H. F., 2014. Good faith obligations under NEC3. [WWW Document]. URL http://www.lexology.com/library/detail.aspx?g=5f583ca9-3c1d-4439-8450-3559ea9538d2.

Williams, T., Williams, M., Ryall, P., 2013. Target cost contracts: Adopting innovative incentive mechanisms to improve the project delivery process. *procs*: 29th Annual ARCOM Conference. 2 - 4 September 2013, Reading, U. K., pp. 759–768.

Wilsoncroft, C., 2009. NEC3: Managing Change. *procs*: ASME 2009 12th International Conference on Environmental Remediation and Radioactive Waste Management. 1 January, 2009, Liverpool, U. K., pp. 25–30.

Wong, P. S. P., Cheung, S. O., 2004. Trust in construction partnering: Views from parties of the partnering dance. *International Journal of Project Management,* 22, 437–446.

Wright, J. N., Fergusson, W., 2009. Benefits of the NEC ECC form of contract: A New Zealand case study. *International Journal of Project Management*, 27(3), 243–249.

Yang, L. S., Yang, H. H., Chen, H. T., Chang, M. F., Chiu, Y., C., Chou, Y. C., Cheng, Y. C., 2012. A study of nurses' job satisfaction: The relationship to professional commitment and friendship networks. *Health,* 4(11), 1098–1105.

Zou, W., Kumaraswamy, M., Chung, J., Wong, J., 2014. Identifying the critical success factors for relationship management in PPP projects. *International Journal of Project Management,* 32(2), 265–274.

Biographical Sketch

Caroline T. W. Chan

Affiliation: Lecturer, Division of Building Science and Technology, City University of Hong Kong

Education: PhD, MSc, BSc

Business Address: Room 5433, Mong Man Wai Building, City University of Hong Kong, Tat Chee Avenue, Kowloon Tong, Hong Kong

Research and Professional Experience: Estimating, Procurement, Cost and contract administration

Professional Appointments: Member of the Hong Kong Institute of Surveyors, Registered Professional Surveyor

Publications from the Last Three Years:

- Chan, C.T.W., 2014. Estimating and Measurement for Simple Building Works in Hong Kong. *Pearson Education Asia Ltd.*, China.
- Chan, C.T.W. & Sher, W., 2014. Exploring AEC education through collaborative learning. Engineering, *Construction and Architectural Management*, 21(5), 532 – 550.
- Chan, C.T.W., 2014. Barriers of implementing BIM in construction industry from the designers' perspective: A Hong Kong experience. *Journal of System and Management Sciences*, 4(2), 24-40.
- Chan, C.T.W., 2013. BIM from Design Stage – Are the Hong Kong designers ready? *procs:* 3rd International Conference on Logistics, Informatics and Service Science. 21 - 24 Aug 2013, Reading, U.K.
- Chan, Eric W.L. & Chan, C.T.W., 2013. Contractor quantity surveyor using ERP system in cost reporting, in: Ahmed, Syed M. (Eds.). *procs*: CITC-VII 2013: Challenges in Innovation, Integration and Collaboration in Construction & Engineering. 19 - 21 Dec 2013, Bangkok, Thailand.

INDEX

E

F

G

H

I

J

K

L

M

N

O

P

Q

R

S

T

U

V

W

Y